R A Bussabarger
Sept 86
desk copy

BASIC CERAMICS

BENJAMIN STEINZOR

BASIC CERAMICS

Photography by Frank Coco
Drawings by Donald Skoro

Prentice-Hall, Inc. Englewood Cliffs, N.J. 07632

Library of Congress Cataloging-in-Publication Data

STEINZOR, BENJAMIN (date)
Basic ceramics.

Bibliography: p.
Includes index.
1. Pottery craft. I. Title.
TT920.S74 1986 738.1 86-8164
ISBN 0-13-058751-6

Editorial/production supervision and interior design by Eva Jaunzems
Cover illustration: Karen Karnes, covered stoneware vessel
Cover design by Photo Plus Art
Manufacturing buyer: Ray Keating
Page layout by Peter J. Ticola, Jr.

Printed in the United States of America

10 9 8 7 6 5 4 3 2 1

0-13-058751-6 01

Prentice-Hall International (UK) Limited, *London*
Prentice-Hall of Australia Pty. Limited, *Sydney*
Prentice-Hall Canada Inc., *Toronto*
Prentice-Hall Hispanoamericana, S.A., *Mexico*
Prentice-Hall of India Private Limited, *New Delhi*
Prentice-Hall of Japan, Inc., *Tokyo*
Prentice-Hall of Southeast Asia Pte. Ltd., *Singapore*
Editora Prentice-Hall do Brasil, Ltda., *Rio de Janeiro*
Whitehall Books Limited, Wellington, *New Zealand*

To Ida

TABLE OF CONTENTS

Color plates follow page 80.

TO THE READER

This book was written for you who are working through your first experiences with clay. It is hoped that it will prove useful whether you are studying pottery in a formal studio course, in more informal circumstances, or are working in your own home or studio. The structure of the book reflects my long experience teaching young adults of college age. Many learned about pottery in a single beginner's class, others advanced with me through a series of studio courses to become potters themselves. I would like to think that the book takes you, the reader, by the hand, just as I would in person, and conducts you through the basic steps, enabling you to acquire both the skills of craftsmanship and visual insights. Above all, I would like the reader to experience with me the satisfaction that comes from a meaningful work successfully completed.

At the start of learning a craft, your needs are special. The subject matter should be basic, and it should be selected for focus rather than wide range. Your skills should be practiced and your perceptions made more acute and truthful. I would like you to learn what to look for. Pottery beginners should acquire experience in more than one shaping method, for the methods reinforce each other in meaning, and variety furnishes a sound basis for the selection of a specialized interest. A good beginner's book also establishes perimeters around each subject to avoid confusing and diverting the learner. At the same time, each subject should be thoroughly explained. The presentation must be methodical and systematic, precise and detailed. All of these special requirements have guided the writing of this book.

A good piece of pottery is the result of a collaboration between the potter's technical knowledge, his sure hand, his sensitive eye, and his sincerity of expression, all culminating in a harmonious work. I firmly believe that excellence of craftsmanship is a fundamental requirement of good pottery, but without sound design and aesthetic meaning, pottery will have little power to persuade. I have therefore organized the contents of each chapter to integrate issues of design, aesthetics, and technical considerations. Chapter Nine discusses visual composition and design at greatest length, but in other chapters, whenever there is a description of a technique, method, or procedure with which visual or tactile properties may be associated, this relationship is fully described.

This book brings together in a single volume the information that beginners need, but it eliminates extraneous information. It is structured so that each chapter contains all the pertinent material under its heading. Thus, for example, the chapter on glaze is lengthy, for it consolidates all background and technical information, methods, and materials. Inevitably, there will be some cross references made by the reader, as no pottery subject can be comprehended in isolation; all to the good. The first chapter constitutes a brief introduction to the whole pottery process. The last is a survey of the history of pottery making. An annotated reading list furnishes titles recommended to assist you as you progress towards higher and more meaningful achievements.

In all sincerity I thank my teachers. Special appreciation goes to Ruth Canfield, my first pottery teacher, who brought me through puppy love with clay into a lasting relationship; Earl V. Pullias, who preached the need for an abiding ethical confirmation in one's life, and who encouraged his students to write "at least" with great clarity and tone; Daniel Rhodes, whose lectures and critiques exemplified truthful observation and incisive perception; and my colleague Duane J. Andersen, partner in teaching pottery for many years at State University College at Buffalo, N.Y.

Benjamin Steinzor

1

A QUICK SKETCH

Nature supplies the materials needed for creating pottery: clay from the earth, fire, and the human hand and heart. Let us imagine early man—he takes clay moist and plastic from the ground, his fingers squeeze and pat, a shape emerges. At first no tools are used, and the modelled objects are permitted to dry hard. The ability of clay to be transformed by the heat of fire is as yet unknown. The response of plastic clay to the hand's pressure is immediate and satisfying. The first objects made very possibly expressed man's sense of existence: his feelings of fear, hope, love, wonder, mystery, and joy, feelings he embodied in fetishes, goddesses, and similar representations. Moist clay was also found to have practical uses for daily living, as when baskets might be lined with soft clay, much as some birds daub their nests with mud.

When the clay dried it became harder and held its shape, but it also became fragile and temporary. How was the process of firing clay invented? Perhaps some finished pieces were by chance left in a hot fire. When the fire had cooled it was discovered that the clay had become hard as stone. It did not dissolve in the rain nor did it break easily or burn. Its condition was permanent, and the finished vessel was treasured for its mundane utility and its sacred expression. Thus began Neolithic man's experience working and firing clay. To this day the basics remain the same: clay, fire, and shaping hands.

Making pottery means making objects out of clay which when fired in a kiln become permanently hard and rock-like. Clay is "refractory," a unique characteristic that enables it to hold up to the heat of the kiln without burning or melting. Pottery traditionally included vessels—both utilitarian and decorative—and tile. Today we open the definition to include a variety of clay forms and diverse statements, many kinds of vessels, but also sculptural work, architectural applications, personal artistic expressions, and fired clay objects without category.

Making pottery is actually only one of several ways man has learned to use ceramic materials and processes. *Ceramics* means using minerals from the earth which are then transformed by heat so hot that it generates its own light; it becomes incandescent. The potter calls this incandescence a red heat or hotter. Glass, too, is a ceramic product, for it is formed when silicate mixtures are melted into a liquid glass by incandescent heat. Other ceramic products are enamel coatings fused to metal (as in bathtubs), plasters and cements, refractories, and structural clay products such as bricks. Pottery is another ceramic process.

Pottery clay is found in the ground. Clays are masses of submicroscopic mineral fragments formed by the natural decomposition of igneous rock. The weathering of the rock has been going on for millions of years, and clays continue even now to be formed in nature. Clay deposits are found almost everywhere. Each deposit is different in its mineral composition from any other, but all clays have a number of common characteristics. The three essential traits held in common are *plasticity*, the ability of the clay to respond easily to the fingers' pressure and to hold that impression; *porosity*, the ability of the clay to retain its shape even after drying; and the ability of clay to become permanently hard and rock-like after it is fired to at least a red heat—sometimes this trait is referred to as *vitrifiability*.

Local clays may be found in a creek bed, an outcropping, an excavation, or even just below the top soil in a back yard. It is not difficult to gather the clay. One simply digs it up with a shovel and takes it away in a pail. However, potters often find it more convenient to buy clay ready to use, prepared by a clay company with just the right amount of water mixed in for optimum plasticity. The processor must first get the clay out of the ground, sometimes from shallow and sometimes from deep mines. Often the mined clay contains contaminating substances such as loose sand and gravel, roots and other organic materials, as well as undesirable metals and minerals, which must be removed. After cleaning, the clay is dried and pulverized into a fine powder, so fine that it is called *air-floated* or *clay flour*. Consumers of very large quantities of clay purchase dry, air-floated clay and add water in their studios or factories to make good, plastic, workable clay. Many studio potters or schools find it more convenient and economical to purchase clay already mixed with the correct proportion of water and packaged in plastic bags ready to use.

We can also make good plastic clay from scratch. In a tub or basin add cold water to dry clay flour or finely powdered dry scraps, then churn and mix the two by hand until the clay is dough-like in consistency. Usually there will be about seventy parts of clay to about thirty parts of water. The process is a bit messy, but there is great pleasure in mixing the soft, muddy mass by hand. When the mix is like a good bread dough or putty, not sticky, the condition is right and the clay can be put away for storage. Clay will remain moist and workable for long periods of time in tightly bound plastic bags or well-covered pails and tubs. It is difficult to restore hardened clay to a good working consistency. Clay does handle much better when it has been aged. Therefore, sufficient material should be prepared and laid away at least several days ahead of the time when the clay will be worked.

Liquid clay, that is, clay that contains about 50 parts of water to about 50 parts of clay, is called *slip*. Slip cannot be used by itself to make pottery. It adapts its shape to the mold into which it has been poured. This is called *slip casting*. Slip is also used as a kind of paint to decorate pottery surfaces and as a glue to fasten two pieces of moist clay together, as in the case of a handle attached to a mug.

Plastic clays respond well to the pressure of fingers, but some more satisfactorily than others. It is easy to tell when a particular clay has reached its ideal plastic or workable state. Roll out a thin coil of clay and bend it around to bring the ends together. If the clay cracks much in bending then it is probably too dry; if the coil is limp and soggy then it needs to be drier. This is also an excellent method for determining the degree of plasticity of a clay you have found in the ground.

The next step in making pottery is always to wedge the clay. Thorough wedging mixes the clay into a homogenous mass and rids the mixture of entrapped air bubbles. Wedging is simple and takes but a few minutes.

Now that we have an adequate supply of good, aged, plastic, well-wedged clay, we can turn to our major work, creating pottery. Pinching clay be-

1-1 *Mixing a small batch of clay.*

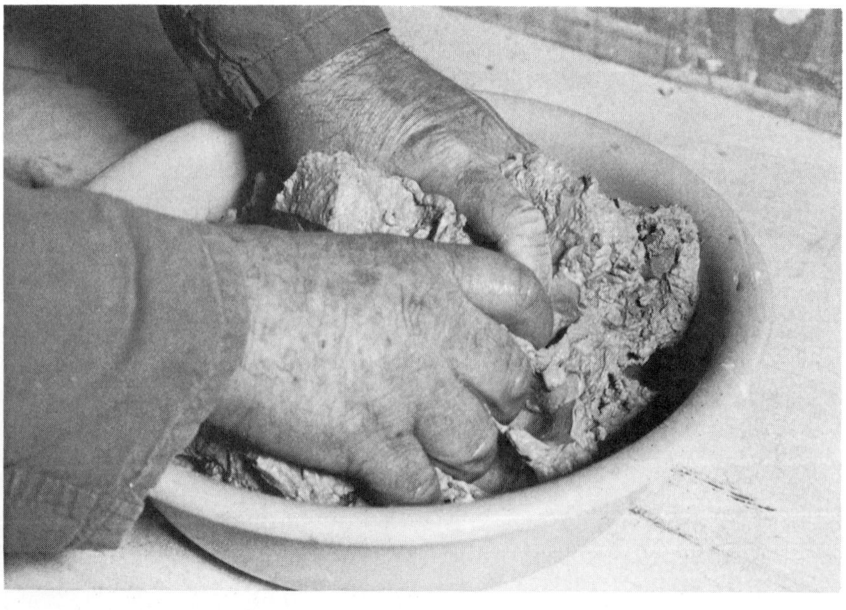

tween the fingers, coil-building, assembling pieces from slabs or sheets of clay, carving and hacking out of solid blocks, piecing together clay pellets, and throwing on the potter's wheel—these are the forming processes. They will be described in succeeding chapters. As a rule, clay works best when the potter has an accurate understanding of its plastic qualities.

Once you have taken clay from its air-tight wrapping, the water in it immediately starts to evaporate and drying begins. As drying takes place the clay shrinks somewhat, generally 5 to 8%. During firing clay will shrink once more. When clay is about half-way dry it is called *leather hard.* There is still some moisture left in the clay at the leather-hard stage, and it still looks moist, but it has lost its plasticity and workability. The leather-hard condition is important to identify, for two pieces of clay can be joined successfully up to, but not drier than, leather hard. There are some decoration techniques and some shaping processes which are best accomplished during the leather-hard stage. Pieces being worked on may be kept from drying by covering them with plastic sheets or plastic bags, or they may be stored in enclosed, air-tight, damp cabinets.

When the piece is completed, it is permitted to dry thoroughly, sometimes in heated drying cabinets but usually out in the open on shelves. When all the water has evaporated the clay is called *bone dry.* If it is really bone dry it should have a powdery look and should feel neither cool nor damp to the touch.

Bone-dry pottery is ready for the first firing in a kiln, the *bisk* or *biscuit firing.* Kilns are insulated structures often shaped like big boxes, inside of which extreme heat can be generated, held, and controlled. Sources of heat for firing may be electricity or burning fuels such as wood and gas. The kiln is somewhat differently constructed and operated for each type of firing. The temperature must be at least high enough to cause incandescence, about 1000°, but in most firings the temperatures obtained will be much higher, in the range from about 1900° to about 2300°.* The dry clay wares, at this stage called *greenware* because they are finished but unfired pieces, are stacked in an empty kiln. The pieces are arranged in the kiln to protect them against damage and to utilize the space of the kiln most efficiently. A good potter stacks the kiln so that it is filled completely. During the firing process certain irreversible physical and chemical changes take place. The wares come out permanently hard and rock-like. The clay color is now its true, fired color. Firing shrinkage has taken place. Most studio and school kilns properly fire bisk in about six to eight hours.

An overnight cooling to room temperature follows, with the door remaining closed after the kiln is turned off.

Some bisk pottery is useful without a glaze; flower pots are good examples. However, pottery is most often glazed. A glaze is a glassy coating melted on the surfaces of the pot. Prior to application, the glaze resembles a creamy or milky liquid mixed with water. Like most glass formulations, glaze is a mixture of silica and other minerals. It is put on the pot by brushing, spraying, pouring, and dipping.

After the piece has been glazed it is once again stacked in the kiln for a second or glaze firing. The primary purpose is to melt the glaze on the pot's surfaces. This firing takes about as long as the bisk firing, and the kiln must again cool down to room temperature. The glaze enhances the visual properties of the form, but its most significant contributions may be in rendering the pot waterproof and giving it a smooth, easily cleaned surface. A sensitive and skillful application of glaze makes the pot complete.

The usual procedure is to biscuit fire and glaze fire separately. This avoids complications and faults which could arise if greenware were glazed without first being bisk fired. Some potters work out special techniques and procedures which enable them to "once fire" their work.

The pottery process outlined in this first chapter is virtually the same now as it has been for many thousands of years—plastic clay is taken from the earth, transformed by fire in a kiln, and given resolution with glaze, all by the hands of the potter.

mined clay→clay flour & water→

plastic clay or slip→leather hard→

bone dry→bisk fire→glazing→glaze fire

*All temperatures are given in degrees Fahrenheit.

2 THE FORMATION, DEPOSIT, AND CHARACTERISTICS OF CLAYS

FORMATION OF CLAYS

All clays, no matter how they may vary when taken from the earth, were at first solid masses of igneous rock. The melted rock, created during the earth's formation, moved up through the crust and gradually cooled and solidified. A number of forces in nature simultaneously and continuously worked on the solid rock to cause its decomposition, and gradually it became clay. This process of decomposition is called *weathering*, and rock is weathered for millions of years in its metamorphosis into clay.

Weathering is accomplished by nature through both physical and chemical attack on the rock. Physical forces break mountainous masses down into smaller and smaller pieces, an aggregate of fragments so small that each can be seen only with the aid of an electron microscope. Climatic changes in temperature cause the minerals in the rock to expand and contract at a rate that is different for each mineral. As a consequence the rock cracks, splits, and separates into ever smaller bits. The action of roots of shrubs, grasses, and trees growing in fissures in the rock contributes to breaking apart the rock. Winds carrying scouring sands blast rock faces and deposit mineral sediment at the surface of the earth. Moving bodies of water and glaciers carry and tumble the pieces together, which are thereby reduced to form a variety of infinitely smaller fragments. These physical forces grind, mix, and to some extent sort the weathered minerals and other materials present.

At the same time, there are chemical reactions that completely change the quality of the original constituent minerals. In the earth, hot gases and acids percolate up through the material, causing further decomposition. Oxygen combines with metals, such as iron, to form metallic oxides. Water washes away some of the soluble minerals.

Out of the former conglomerate of minerals a new composition takes shape, called feldspar, a mineral that is an arrangement of alumina, silica, potassium, sodium, and calcium in a variety of combinations. Feldspars are of value for a number of uses, among them as a constituent for pottery bodies and glazes. Feldspars are continually subject to weathering in the earth.

Water gradually combines with alumina and silica in feldspar. Potassium and sodium are washed out. A new mineral structure has been formed, called *kaolinite*, a hydrous silicate of alumina. This critical reaction of water with feldspathic rock is called *kaolinization*. Kaolinite is the name for the pure clay substance in the mass. In truth, there is no such pure clay substance to be found in nature, but it is that which imparts to the whole clay mass its characterizing qualities.

Such are the forces of nature that there is always present in the clay mixture other minerals and materials which we call *accessory minerals* and *contaminants*, or impurities. Clays vary widely in the degree of their purity. The extent to which the sediment in a particular clay deposit has been weathered and the nature of the contaminants give a clay its individual composition and behavior.

Some of the more commonly found impurities in clay are sand, feldspar, and iron. The size of the sand particles may be anywhere from coarse and gritty to the particle size of the clay itself. Coarser sand reduces workability in the clay, but it has the advantage of permitting water to evaporate more easily. The very finest sand or quartz promotes greater hardening of the pottery during firing, because the silica begins to melt into a fused matrix.

Feldspathic substance in clay also promotes hardening during firing because the feldspars melt, imparting glassiness within the clay structure. The ingredients that advance glassiness or fusion in firing are called *fluxes.* Hardness in the fired clay results from the glassy state that holds clay particles in tight bond.

Iron appears in a good many clays, particularly those that have been deposited near the earth's surface. Its most notable influence on the behavior of clays is on color in the unfired and the fired states. A clay high in iron, perhaps as much as 5% or more, will have a characteristic brick-red fired color, as in a flower pot. A small amount of iron may color the clay salmon or buff. Some clays may also contain traces of calcium, potassium, sodium, and manganese.

CHARACTERISTICS OF CLAYS

All clays contain a sufficient quantity of hydrous silicate of alumina or kaolinite to impart to the clay mass the three vital characteristics by which clay can be identified: plasticity, porosity, and vitrifiability.

Plasticity By plasticity we mean the ability of clay to take a shape from the hand's pressure and to retain that shape. Clay does not spring back as does rubber. The platelet or slate shape of the clay particle itself provides probably the most fundamental explanation of the clay's plasticity. When a wet film penetrates between these multitudes of flat particles, they slide sideways in relation to each other and resist pulling apart or falling away. The minuscule size of the clay particle also affects plasticity. Finer grained clays are more workable or plastic than coarser grained clays.

Aging a supply of clay also improves plasticity. The older the clay the more thoroughly wet are the spaces between the flat particles. As a clay ages a dark, moist mass and organic or mold life is apt to grow, which acts as a lubricant for the particles to slide against each other. Also there will be some degree of continued decomposition during long storage. An extremely plastic clay may be referred to as a *fat clay,* while a much less workable material may be called *short.*

Porosity Porosity is the quality that permits the water, added to the dry flour by the potter, to evaporate while letting the pot retain its shape. The added water is called *the water of plasticity.* Pores or spaces among the clay particles hold the water of plasticity, with coarser grained clays drying more swiftly because of larger pores. (As a consequence of their geological histories, some clays are coarser grained and some are finer grained.) As the pot

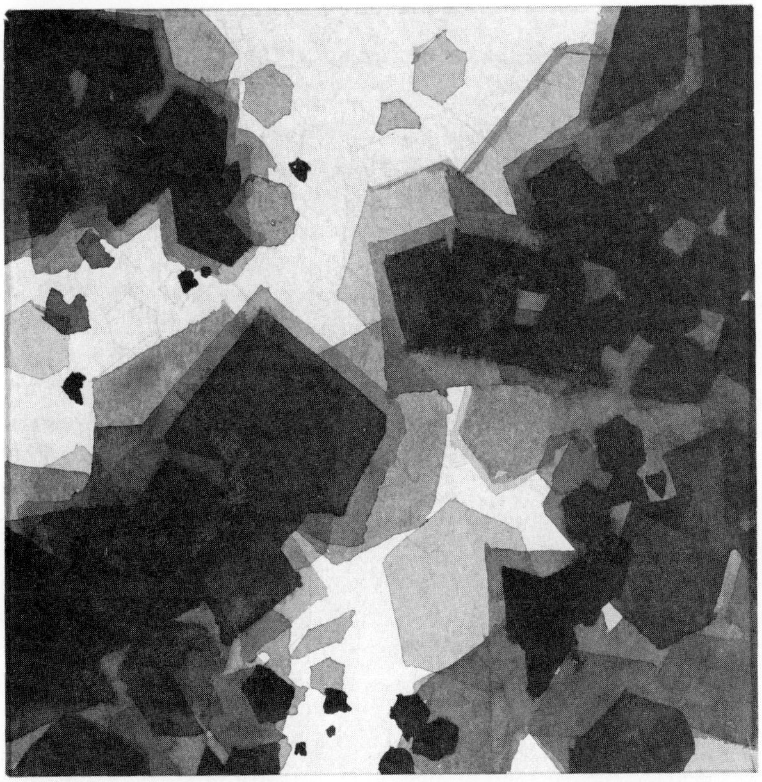

2-1 *Kaolinite crystals have platelet shapes.*

dries, the water evaporates from the pores, the clay particles draw closer together, and drying shrinkage takes place.

There is more shrinkage in clays that are finer grained. Much of this shrinkage takes place in the early stages of evaporation, with less shrinkage after the leather-hard stage. During their formation and deposit, many clay deposits have had mixed in with them some sand or quartz and other varieties of non-plastic sediment. Unlike clay particles, these particles are not slate shaped; they are more round, and cause larger pores to permeate the clay mass. Clay with a smaller proportion of pure clay substance, but with more non-plastic material, dries more quickly and with less shrinkage.

Vitrifiability Vitrifiability is the characteristic of clay that renders it permanently hard and rock-like after it has been fired in a kiln. During firing, fluxes, such as feldspathic minerals, begin to melt together with the pure clay particles and other minerals present. A glassy bond is formed. As this fusion occurs, the particles draw closer together and fired shrinkage takes place. The clay becomes hard as a consequence of developing vitrification. The higher the fire the more melting and shrinkage there is. Ultimately, all clays would become hard enough to be waterproof even without glaze if fired high enough, but the practice most often is to stop firing before this takes place. Should the firing temperature be too high, the clay would, in most cases, melt so much that the piece would lose its shape; the pot would begin to droop and sag. The ideal state is termed *maturation*, that is, the clay is fired as high as it will go without deformation.

TYPES OF CLAYS

The standard method for potters to categorize or classify clays is by their location in the earth, where they have been deposited by nature. Clays which are found at or near the site of the parent rock, the rock masses from which they were formed, are called *primary* or *residual clays. Secondary, transported*, or *sedimentary clays* are those which have been deposited after being removed by nature from their original locations.

Primary (Residual) Clays

Residual clays are often found overlying their parent rock, and they grade down from fully formed clays at the surface to the original rock at depth. Feldspars may be found intermixed. The feldspars are gathered for use separately in pottery bodies and glazes. The residual clay of importance to potters is *kaolin*. "China clay" is another term for "kaolin." The word "kaolin" is derived from the Chinese "Kaoling," meaning high ridge. Over 1000 years ago the Chinese used this expression to refer to the clay material taken from ridges, from which they made porcelain.

Kaolin is the purest of clays. That is, it contains the largest proportion of the clay mineral kaolinite. Its formula most nearly approximates the formula for kaolinite, and the formula is used interchangeably for both. Kaolinite has the theoretical formula $Al_2O_3 \cdot 2SiO_2 \cdot 2H_2O$. This means that there is present about 40% alumina, 45% silica, and 15% water. As a convenience, the formula is used to represent all clays.

Kaolin fires white; it contains virtually no iron or rutile to discolor it. Because of its geological history it is more coarse grained and, therefore, relatively low in plasticity. Kaolin is a particularly refractory clay; that is, it has a high resistance to heat, due to its large particle size and the absence of fluxes in its composition. It does not become glassy at an early firing temperature. As a result of its low plasticity and high refractoriness, it is not practical to use kaolin alone to make pottery, since it does not work easily or fire conveniently. Rather, more plastic clays and fluxes are added to make a mix or clay body that is suitable for pottery making. Kaolin is also often a component of glaze recipes, for it contains alumina and silica for the glaze. In the United States the major deposits are in North Carolina. As is characteristic of all clays, each individual deposit varies somewhat from every other. That is why clays are identified by their own brand names, such as E.P.K. Kaolin or Kamec Kaolin.

Secondary (Sedimentary) Clays

Secondary or sedimentary clays are carried along as sediment by moving bodies of water, glaciers, winds, or the pull of gravity. While the particles are moving along they are simultaneously ground into smaller sizes; they are mixed with each other and mixed with other minerals and materials from the surrounding environment; they are sorted as to size; and they are deposited.

Secondary kaolins are kaolins that have been transported from a residual deposit. They are quite similar to the residual variety but are somewhat finer grained and consequently more plastic. Many secondary kaolins contain a trace of iron and do not fire quite as white. They tend to be less refractory—slightly less heat is required to mature them in the fire. The major sources of these clays in the United States are South Carolina, Georgia, and Florida.

Additional types of secondary clays are ball clays, fire clays, stoneware clays, and dark-firing clays.

Ball clays fire a light ivory to buff color. They are extremely fine grained and shrink very much

TYPES OF CLAY

Primary	Secondary
kaolin	kaolin
	ball clay
	fire clay
	stoneware clay
	dark-firing clay

in drying. Ball clays are never used by themselves to make pottery, but because they are so plastic they are added to clay bodies to increase the plasticity of the mixture. The most important deposits are found in Kentucky and Tennessee.

Fire clays are noted for their extraordinary resistance to both melting and heat stress, due to the large grain size and the presence of alumina and silica as accessory minerals. In the ceramics industry fire clays are components in the manufacture of refractory products, parts for furnaces, kilns, and the like. Potters sometimes add fire clay to their clay bodies for the grain textures and sturdiness. The grainy texture or hand is called *tooth.* These clays are found in many places, among them Missouri and Ohio.

Many *stoneware clays* possess sufficient plasticity, good color, and texture so that they may be used alone to make good quality pottery. They usually occur as red-brown or grey firing deposits; the color results from the presence in the deposits of small amounts of iron, rutile, and traces of manganese. Rutile and feldspathic material in the stoneware act as fluxes, enabling firing to maturity somewhere between 2100° and 2350°, a range high enough to produce a very hard, nearly impervious, and durable ware. The pottery produced from this clay is generally named stoneware. Deposits are found in many places, notably Ohio, New Jersey, Illinois, and Pennsylvania.

The most common clays are called *dark-firing* or *earthenware clays*, and sometimes *terra-cotta clays.* They occur as deposits or shales very close to the earth's surface, a matter of great annoyance to gardeners. These clays are usually quite plastic and contain large amounts of iron, firing buff to red-brown colors. Bricks, drain tile, and flower pots are among the common commercial products. Of all the clays, these require the lowest heat to mature them, in the neighborhood of 1800°–2000°. Because of their ready accessibility and the convenient low firing permitted, probably more pottery has been made from these clays than any other variety.

Bentonite

There is still another material, bentonite, which is clay-like but not exactly a clay mineral. Bentonite is the result of the decomposition or weathering of volcanic ash. Therefore, its particle size is even finer than any clay, so fine that its use is limited. It may be added in fractional amounts to clay and glaze mixtures for the high plasticity it contributes. For example, 1% to 2% of the total may be added to a pottery body; that small amount noticeably improves plasticity. The very fine grains of bentonite in a glaze help the coarser mineral particles in the watery mixture stay afloat and not settle to the bottom.

In sum, many students of pottery will recognize the following definition of clay learned from respected teachers.

Clay is a natural aggregate of fine grained mineral fragments in various stages of alteration. Sometimes clay consists of a single substance but it is usually made up of different kinds. However, it should contain as its essential ingredient some form of the mineral hydrous silicate of alumina. Clay should be plastic when mixed with water, a property which enables it to be molded into any shape, which shape is retained when dried out, and it should become hard and rock-like when fired to a red heat or higher.

3

FINDING AND PREPARING LOCAL CLAYS

Local, native, common, or dark-firing clays are deposited widely and found in many places. The deposits are not always high enough in quality to warrant commercial clay mining; they may not even meet the modest needs of a pottery studio. Perhaps the deposit is a small one, it is not uniform, or it has too much sand, organic material, and contaminants, such as calcium and sulfur, intermixed. These ingredients adversely affect the plastic, drying, and firing behavior of the clay. In sufficiently large quantities interspersed throughout the mass, they prevent the clay from functioning well in plasticity, in drying, and in firing. Nevertheless, there are times and places when a small amount of usable local clay is easily discovered and can be put to use by an individual potter. Then, there is great satisfaction derived from experiencing the pottery process from its very beginnings to its conclusion: to dig perhaps a couple of buckets full of clay, make it ready for use, and then shape the clay, fire, and glaze the pot. It is not often in modern times that we have the opportunity to participate at all levels in our work. When we can, the fulfillment is deep.

Finding Local Clays

Native clays may be seen in the bank or bed of a creek or stream or at the site of an excavation. Sometimes layers of clay are found uncovered, exposed by long erosion. The clay condition can be distinguished by its texture and color from surrounding loam, sand, and gravel. If the clay is dry, it will appear solid and pitted rather than loose and washed away. The color is usually tan, rusty, or grey-green, evidence of the presence of iron in one form or another. Most of these local clays have a large amount of iron mixed in with them, perhaps as much as 5% to 10% of the total. A shiny or slippery surface gives the clue that the moist material will hold together and not crack when a handful is rolled into a coil and bent around into a doughnut shape. Let the clay doughnut dry. Should it retain its shape, uncrumbled, that is even stronger evidence that it is clay. If it is convenient to take the specimen to a kiln and fire it to about 1850°, we would have the final test as to its value as a working material. The piece should come out of the kiln hard as rock, somewhat smaller than its original size, and probably some shade of iron in color. It may be interesting to test the clay thoroughly, following the procedures described in the next chapter. The results of these standard clay tests tell a great deal about a clay's characteristics.

These small deposits of clay sometimes exist in sufficient uniformity and plasticity that pottery can be made directly from the clay as it is taken from the earth. In other locations the clay might be a bit too sandy in texture, crumbly, uneven in consistency—not quite usable as is, but good if refined. Then the clay can be made usable in a series of simple steps.

A shovel and some pails will be needed for the digging. The dug clay should then be separated into pieces, nut size or smaller, and spread on a clean surface to dry. If the weather is right, leave them out in the sun. Drying may take a day or two. When bone dry, the bits should be smashed and pulverized. Use rolling pins, heavy sticks, boards, or mallets.

3-1 *Testing for plasticity.*

Refining Local Clays

Have ready several pails or basins half full of water. Into each the dry clay is slowly poured until it piles up in a cone with the point just above the water level. Allow the clay to soak or slake for a day or two for the most thorough wetting. Then stir the mixture by hand or with a flat paddle until a smooth, creamy slurry results. Soft lumps in the slurry should be broken up and dispersed by hand.

Foreign material in the liquid clay, such as gritty sand and traces of roots, must be removed by screening. A screen at least the size of a 40 mesh should be used. (The number indicates the count of openings in a linear inch. To compare, most window screens have a mesh of about 14.) The finer the mesh used, the more of the contaminating material is removed. Special screens or screen cloth from which you can make your own sieve may be purchased from a pottery-supply company. The clay slurry is poured through the screen into an empty pail or basin, but first the screen should be wet so that its pores will not clog. Should the slurry not flow easily through the screen, add more water and stir thoroughly to thin the liquid.

3-2 *Pour dry clay into a container of water.*

3-3 *Screening clay slurry.*

Preparing Local Clays

After the clay slurry has been screened, it must be converted into a plastic mass. Liquid clay, or *slip*, contains about 50% water to about 50% clay material. Good, plastic clay contains about 70% clay to about 30% water. To turn liquid clay into a solid state, the excess water must be removed. A

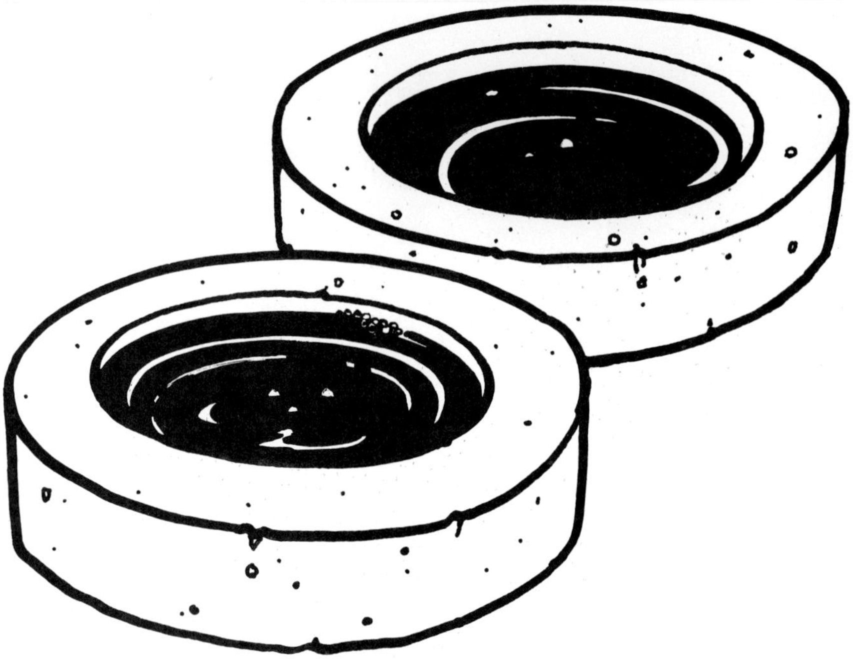

3-4 *Plaster drying bowls transform clay slurry into plastic clay.*

pottery factory or large studio will have machines to squeeze the water out of the slurry; these are called *filter presses.*

Where small amounts of clay are being worked by a potter, there are practical and simple means for removing excess water. Out in the country, the clean, screened slurry can be poured onto a large, flat rock to dry in the sun. Water will evaporate slowly in the warm air. In the studio pour the slip onto canvas-covered tables or large, thick boards. Some of the water is absorbed by the porous material the slip rests on. Thick plaster of Paris drying bowls or plaster slabs may be made for the purpose, as described in Chapter 17. The plaster effectively absorbs water; gradually the clay solidifies. In time, perhaps overnight or in a day or two, the clay will have assumed its proper plastic condition. It will have the consistency of a good bread dough, putty-like, with just the slightest degree of elasticity or rubberiness.

Clay in a good workable state is not sticky to the fingers, but it will leave the hands moist. The mass of clay should now be put away to age. Clay works better after aging several days but even more satisfactorily after several weeks. Store the clay in an enclosed space, such as a plastic bag, plastic or galvanized garbage pail, or a large crock. All wrappings and containers must be tight to prevent evaporation and drying.

By much the same method, clay scraps from around the studio may be turned back into good, plastic material. Pot trimmings, odd pieces, whole pots gone bad, and slurry from the potter's wheel may simply be put into a large pail or basin with water. Most pottery studios have a "slops" pail in place for this purpose. When the pail is filled, the process described above is repeated. First screen the slurry, then remove the excess water by drying on a large absorbent surface or in plaster drying bowls.

4

COMPOSING POTTERY BODIES

A clay may not be in every way suitable for making pots. It may be short in plasticity, or its texture may be uninteresting. The clay may dry too slowly and shrink too much; it might have poor dry strength or an unexciting color; it may be incompatible with available glazes; or it may too easily develop firing faults. For these and other reasons a potter might rather formulate an original mixture of clays and appropriate minerals to overcome real faults and perceived weaknesses. In this new mixture there will be deliberately blended ingredients chosen by the potter for desired unfired and fired characteristics. The mixture is very much like an alloy in metallurgy. It is called a *clay body* or *pottery body*, to differentiate it from a single, naturally occurring clay.

Each of the materials in a body formulation affects the whole behavior of the body. If the ingredient contributes to greater plasticity we call it a *plastic agent*. A *filler* is one that aids the clay in drying more rapidly, with less shrinkage, cracking, and warping. A *flux* helps the pot become mature when fired at a lower temperature than otherwise.

PLASTIC AGENTS

All clays are, to varying degrees, plastic agents. To increase plasticity, it may be desirable to use more than one clay in a body. A fat or very plastic clay, such as ball clay, greatly enhances workability, but it may also have adverse effects. The fat ball clay also increases the shrinkage during drying and the possibility of warping, and at the same time slows down drying time. A short clay such as kaolin, or a coarse grained clay such as a fire clay might be introduced to adjust the balance of grain sizes. Some bodies may contain as many as three different clays, each making its unique contribution to the mixture. To change the balance of clays is also to alter somewhat the maturing temperature of the body. Stoneware clays and ball clays would lower the firing temperature compared to kaolins and fire clays, which would raise the temperature. The most suitable compromises for good plasticity, high dry strength, acceptable shrinkage, and interesting color and texture are arrived at by trial and test.

FILLERS

Any non-plastic ingredient in the body may be considered a filler material. The particle shape of fillers is usually round or grit shaped, and the particle size is larger than the minuscule assortment of clay particles. As a consequence, pores or spaces among the fragments are more spacious and permit water to evaporate more easily; shrinkage is reduced because the size and shape of the filler particles do not allow the clay particles to draw closely together. The major fillers are potter's flint, sand, and grog. Flint is ground quartz (silica), and sand is merely a more gritty form of silica. Grog is clay that has been fired and crushed into a sediment. It is available in a number of meshes.

Fluxes and short clays also behave partially as fillers, but this aspect should be considered a by-product of their other, more essential, functions in the composition. At the same time that a filler permits more rapid drying with less shrinkage, cracking, and warping, it reduces workability or plasticity. Fillers also have the disadvantage of decreasing dry strength. The potter puts the clay body through a series of tests and trials, for the purpose of a satisfactory balance of non-plastic fillers with plastic clays.

FLUXES

Clays become hard during firing because the heat of the kiln is high enough to cause one or more of the minerals in the blend to begin to melt into a glassy phase. The higher the heat, the more melting takes place. The more substances in the clay that can melt or fuse at a given temperature, the harder the clay becomes. Those minerals that promote melting at a lower temperature are called *fluxes*. The major natural fluxes are the feldspars, Cornwall stone, nepheline syenite, and talc. Feldspar responds very actively to the effect of the kiln's heat, particularly above 2150°. Cornwall stone, imported from England, acts like feldspar but requires a higher heat for it to go into a melt. Nepheline syenite and talc are fluxes that are more often used in body compositions fired at lower temperatures, such as earthenware.

Body frits are man-made fluxes, resembling glasses in their composition. Frits are almost always used in low-temperature wares.

Too much frit or too reactive a flux is like too much of a good thing. During firing the melting may become so advanced that the pot sags and droops; after firing and upon cooling the clay might become brittle and shatter easily. If a low-fire clay is one of the ingredients in a high-fire body, less flux may be incorporated, for the composition of the low-fire clay causes it to act as a flux at the higher temperature. The major criterion for selecting which flux to use and how much to put in a pottery body is how successfully the pottery can be fired to its ultimate hardness without melting or warping. This ideal condition in fired ware is called *maturity*.

COLOR

Minerals are sometimes added to a clay or body simply to enhance the aesthetic qualities of the material, for improved color and texture. Color can be changed by introducing a small quantity of any of the various metallic oxides. The potter has more precise control over color results if a light or white firing composition is the base. Many of the available clays have at least a trace of iron, a consequence of weathering and deposit. Often the iron modifies the effect of the added colorant. However, should a strong iron color be desired, ranging from buff to rusty-red, additional iron must be introduced. It is best to add iron in the form of a dark-burning clay, such as Redart.

Colorants often behave as fluxes when added even in very small amounts to a clay body. This influence should be anticipated by adjusting the proportion of flux or the firing temperature. The studio potter has available commercially prepared colorant mixtures in a very wide range of hues. These pigments are called *body stains*. They are uniform and predictable.

Colorant Oxides

Many potters prefer to use the common metallic oxides as colorants. They are considerably less expensive than commercial stains. Metallic oxides offer any potter a wide opportunity for rich exploration of color. They may be used singly or in combination. If more than one is incorporated in a body the amount of each should be reduced. True body color develops in the second, or glaze, fire. The firing mode, whether oxidation or reduction (see Chapter 11), also determines dramatically the interaction of colorant, clay, and glaze. The most frequently used oxides are listed below, together with suggested percentage ranges, and results expected in electric kiln and oxidation firing:

cobalt oxide	½–1%	blue
copper oxide	1–4%	green
iron oxide	1–5%	tan to rust
manganese oxide	1–4%	earthy lavender
iron chromate	1–4%	grey
nickel oxide	1–4%	grey greens
ilmenite (granular)	1–2%	speckles
manganese (granular)	1–2%	speckles

Color in the clay is intensified in direct proportion to how much colorant is introduced. (A thorough discussion of color is in Chapter 9).

TEXTURE

Any coarse grained or non-plastic material increases the grittiness or tooth of the clay in direct relation to both the grain size of the material and the quantity added. One might introduce into a clay or a body a small amount of fine or medium grog, in the neighborhood of 10% to 20%, about the same amount as found in a coarse grained fire clay. The result would be a material that is rougher in grain under the fingers and, after firing, coarser in texture under the glaze. Moreover, particularly to handbuilding or sculpture clays, any one of a variety of substances may be added that will subsequently burn away during the firing. Coffee grounds, rice grains, bread crumbs, rope fibers, sawdust, and wood chips are some of the many possibilities. When these substances burn away small holes shaped like the now-absent grains remain permeating the clay structure. The clay turns sponge-like in texture. During the wet, workable stage and during drying these materials function as fillers.

BODY RECIPES

The listing of ingredients for a body composition resembles a food recipe. Several examples are given at the bottom of this page. The "cone" designation is a measurement of the firing temperature (see Chapter 12).

STANDARD CLAY TESTS

Pottery bodies are formulated to meet the needs of plasticity, porosity, and vitrifiability, and these criteria are verified by a series of trials and tests. Materials incorporated in a body should not adversely disturb these fundamental requirements. Ingredients are chosen for the good characteristics they can contribute and for how they might positively complement each other. Although measurements for glazes should be done with exacting precision, in the small pottery studio a certain amount of flexibility is permitted in body formulation. The pottery body composition may be juggled and adjusted until it satisfies working, drying, firing, and aesthetic requirements.

Most studio pottery bodies fall neatly within one of three categories of pottery wares: earthenware, stoneware, and porcelain. The temperature range within which each is fired is a significant element in its definition. Most earthenware is fired at some point 1850–2100°. Stoneware's firing is generally 2200–2350°. Porcelain, the hardest ware, requires the highest fire: 2350–2500°.

A few tests can be conducted in the studio to disclose how a recipe would work out. Test results indicate the alterations necessary for fine tuning the recipe. The important tests are for the amount of water of plasticity, how much shrinkage takes place in drying and firing, and the relative amount of water absorbed by the fired clay.

How much water is needed to make the mixture properly plastic is a good clue as to the actual workability of the clay. A finer grained clay or body mixture takes up more water, dries more slowly, and shrinks more. Water of plasticity much in excess of 30% is a flag for caution.

The test for drying shrinkage indicates whether the clay body will behave safely in drying. Excess shrinkage is often accompanied by warping and cracking. If we know the amount of dry shrinkage we can more productively plan finished sizes and more successfully fit parts such as lids.

Shrinkage during firing is as predictable as drying shrinkage. Firing shrinkage gives us information we need for proper sizing and fit of parts. It is a vital clue as to the correct maturing temperature for the clay body. The higher a piece is fired, the more it shrinks. Therefore, it is advisable to make a number of tests at progressively higher firings to trace the advance of hardening in the clay mass.

A body sample should also be tested for how much water it is capable of absorbing in the fired state. As clay is fired to ever higher temperatures, more glassiness is developed within the clay structure; as a consequence, there is greater shrinkage when the glassiness causes the mineral particles to draw more closely together. A clay structure that

CONE 04	
Earthenware	
kaolin	30%
ball clay	45%
body frit	5%
talc	10%
flint	10%
	100%

CONE 04	
Ovenware	
fire clay	50%
EPK kaolin	20%
fine grog	20%
talc	10%
	100%

CONE 4–6	
Stoneware	
Goldart	20%
fire clay	15%
EPK kaolin	20%
ball clay	20%
nepheline syenite	20%
flint	5%
	100%

CONE 9	
Stoneware	
Goldart	60%
fire clay	20%
ball clay	12%
F-4 spar	6%
Redart	2%
	100%

CONE 10	
Porcelain	
kaolin	35%
ball clay	20%
feldspar	25%
flint	20%
	100%

is more dense is less able to absorb water. Thus, the test for water of absorption in the fired state is essentially a test for fired hardness: More absorption is an indicator of softer clay while less absorption is evidence of hardness. Earthenware may be expected to have an absorption rate of 5% to 10%. Stoneware characteristically absorbs 2% or less of its own weight in water. Porcelain, by definition, has zero absorption; it is completely dense.

The tests are often used as a component part of the body-formulation process. They are also valuable to gain an understanding of the behavior of a single, naturally occurring clay. Just a few tools are required:

a glass plate, sheet of plastic, or formica tabletop,
a metric scale (balance),
a 500-cubic-centimeter graduate,
a ruler with centimeter markings.

DETERMINE WATER OF PLASTICITY

1. The clay or body should be pulverized and thoroughly mixed.
2. Weigh a 500-gram sample. Place the powder on glass slab or formica table top.
3. Have a 500-cc. graduate of water filled to the 500-cc. mark.
4. Slowly add the water to the clay. Keep a count of the water used.
5. As water is added, mix it with the clay by hand.
6. Stop adding water when the ideal plasticity has been reached.
7. Use the following formula to determine the percentage of water of plasticity:

$$\frac{\text{cc. of water used}}{\text{weight of dry clay}} \times 100$$

DETERMINE PERCENTAGE OF DRY SHRINKAGE

1. Have ready in the plastic state a small sample of clay from the above test.
2. Form two bars $\frac{1}{4}''$ thick × $1\frac{1}{2}''$ wide × $6''$ long.
3. With a needle or knife inscribe a line along the length of each bar. Mark off a ten-centimeter length on each line.
4. When the bars are bone dry, measure the dry length of the line with the ruler.
5. Use the following formula to determine the percentage of dry shrinkage:

$$\frac{\text{plastic length} - \text{dry length}}{\text{plastic length}} \times 100$$

DETERMINE PERCENTAGE OF FIRED SHRINKAGE

1. Fire two bars from the preceding test to the desired temperature.
2. Measure the fired length of the line with the ruler.
3. Use the following formula to determine percentage of fired shrinkage:

4-1 *Plastic clay shrinkage bars. The marks are ten centimeters apart.*

$$\frac{\text{dry length} - \text{fired length}}{\text{dry length}} \times 100$$

DETERMINE PERCENTAGE OF WATER OF ABSORPTION

1. Bars should be fired to the temperature desired.
2. Bars should be completely dry.
3. Weigh the bars on a metric balance.
4. In a pan, soak the bars, covered with water for 24 hours, for complete saturation.
5. Wipe surface moisture from the bars, using a cloth or towel.
6. Weigh the water-saturated bars.
7. Use the following formula to determine the percentage of water of absorption:

$$\frac{\text{soaked weight} - \text{dry weight}}{\text{dry weight}} \times 100$$

SLUMPING

To determine the resistance of a clay to a given heat:

1. Place a dry test bar over two one-inch shelf posts, forming a bridge with them.
2. Set the bridge in the kiln and fire it to the desired temperature.
3. If the test bar slumps in the fire, it indicates the degree of excess heat by the extent of slumping. If the test bar comes through straight, the clay has either received insufficient heat, or it has been fired correctly. The results of the shrinkage and absorption tests will indicate which of these applies.

5 GETTING CLAYS AND POTTERY BODIES READY

BUYING CLAYS

Plastic clay can be purchased in quantities that are easy to handle, tightly wrapped in plastic bags so that the clay keeps well over long periods. Most pottery-material suppliers carry a variety of plastic clays and bodies, covering a wide range of firing temperatures, colors, textures, and purposes. From this broad palette, the aesthetic needs of most potters may be satisfied. Choose a clay that is most compatible with the firing range of the kiln available, the glazes to be used, and the kind of work anticipated. Purchased, prepared clay is probably the most convenient form of clay supply for the average studio and classroom. All of the time-consuming, technical, and messy work has already been done. Often, purchased clay has been wedged in the factory so that the process need not be repeated in the studio.

Nevertheless, there are some limitations imposed when a prepared clay arrives in the studio. The most obvious is that commercially prepared clay is more expensive; the cost includes preparation labor plus the added freight of water. Perhaps the most important disadvantage is that the potter may not be able to find a material precisely suited to his temperament or style and visual and technical requirements. Or the potter may want merely to alter one of the body qualities, changing it slightly. The drawback is that body materials cannot successfully be incorporated with plastic clay. It is difficult to blend in most textural agents and fluxes, and it is impractical if not impossible to mix dry colorants with a moist mass. Blending ingredients is a simple task, however, when all components are powders.

When the clay is more completely under the potter's control, the potential for good work is enhanced. There is great satisfaction and a sense of creative accomplishment when the potter's hand is in the whole effort, commencing with making clay.

MIXING CLAY BODIES

All body materials, such as clays, feldspars, flint, and colorant oxides, are available for purchase in pulverized form. After being mined, these ingredients are ground to 200 mesh or finer. Most clays are so finely ground they are called *air-floated* or *clay flour.* When coarse grained textures, such as from grog and fire clay, are desired, there are larger mesh sizes available. Body materials can be purchased in any amount. In their dry condition they keep well indefinitely.

It is necessary to have a good body recipe to start with. Recipes are available in many pottery books and magazines, as well as from friends and associates who are potters. A few sample clay body recipes are given in the previous chapter. Of course, an original composition can be formulated from scratch, using the information developed in the preceding chapters.

As noted above (p. 13), pottery bodies and usable clays generally fall into one of three categories, depending on how high the wares must be fired to become mature. In increasing levels of firing temperature, they are earthenware, stoneware, and true porcelain. Within each type many recipes are possible, attributable to the complex of relationships influenced by the kinds and quantities of ingredients; the choices of maturing temperature, plasticity, porosity, color, and texture; and how the wares are to be used. (A full discussion follows in Chapter 16).

It is always prudent first to mix a small sample and test the recipe for plasticity, shrinkage, and absorption, as described in Chapter 4.

Preparing one's own clay body is satisfying but messy work. Some think the messier the more satisfying. Mix as large a batch at one time as studio conditions permit. The result is a good store of clay to age. Weigh the ingredients on a kitchen scale, a bathroom scale, or a studio platform scale, compensating for the weight of the materials container. Place the weighed ingredients all together into a sufficiently spacious mixing pail or basin, a garbage pail or trough.

Some potters enjoy mixing the body in a cement trough or sturdy plastic wading pool. After all the dry ingredients are in the trough, water is gradually added while the potter stomps through the clay with bare feet until a good plastic mass results, much as old time wine makers trod their grapes.

5-1 *Mixing a clay body by hand.*

Another method is to sift the dry ingredients into a large pail of water, through the fingers or a coarse mesh screen. Pour the dry powders slowly into the water, to avoid lumping. There should be enough water in the pail to take all the powder up into a slurry. Give the mix time for complete saturation. Blend the slurry thoroughly. There are electric mixers, but good will, a little elbow grease, and a wooden paddle will do nicely. Have all of the parts dispersed well into a creamy, uniform slurry.

The excess water must now be extracted. Ladle the slurry out into plaster of Paris drying bowls, on plaster drying slabs, canvas-covered table tops, large wooden boards, or even many thicknesses of newspaper. When the extra water evaporates or is absorbed, the slurry is transformed into cakes of workable, plastic clay. This method makes very good clay, and works just as well for preparing moist clay from clay flour.

Very small batches of clay can be prepared by combining the clay flour and water inside a plastic bag, such as a food storage bag, then kneading and mashing the bag from the outside until the proper consistency results. Add flour or water to the contents as required for good plasticity.

A tub or basin filled about one third full of water works well. Slowly sift clay flour into the water until it fills the tub to at least an inch above the water level and allow the clay to settle overnight. If the clay is still too dry sprinkle it with water; if the clay is too soft spread some more flour across the surface, and allow the mix to sit a while longer.

A final method for preparing small batches of clay is to combine water and clay flour in a small basin, and simply mash, churn, and knead the contents by hand until it is thoroughly mixed. Add

5-2 *Clay flour combining with water.*

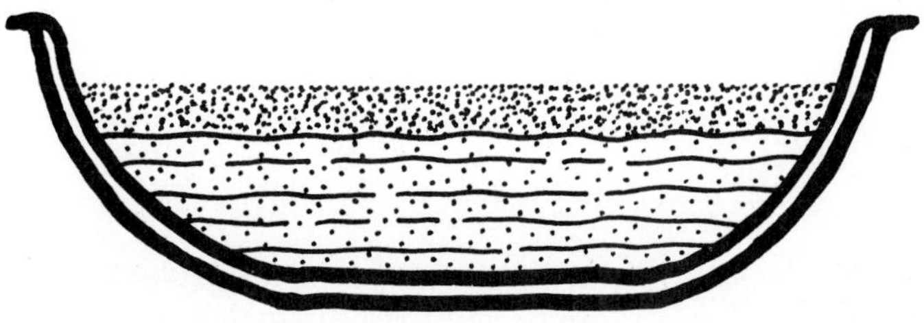

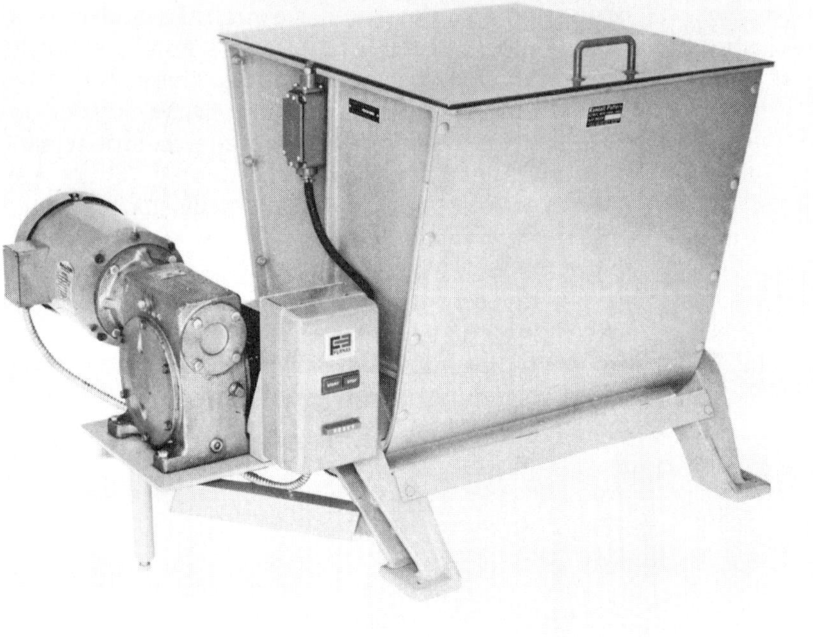

5-3 *Randall Clay Mixer, Model CM.*

Randall Pottery, Alfred, NY.

5-4 *Shimpo Pug Mill.*

Shimpo-West, Inc., City of Industry, CA.

either water or flour as needed to adjust the mix for good plasticity.

There are electric machines of different types for all of this but they are generally only practical in large studios.

KNEADING

Once the body or clay has been made into a plastic mass, the clay should be kneaded in small batches for a short while, as if it were bread dough, on a wooden table top or a large plaster slab. Kneading transforms the loose conglomerate into a cohesive, responsive, tensile, slightly rubbery, and resilient whole. The clay may now be put away for aging and storage. Use tightly covered basins or pails, or wrap the clay securely in plastic bags. Clay aged a week or more works noticeably better and responds much more sensitively to the potter's fingers than clay aged for a shorter period of time.

WEDGING

Wedging is the final step to prepare the clay. It renders the mass completely uniform in texture and devoid of air bubbles, which are almost always present in unwedged clay. Air pockets must be removed because air trapped in the pot's wall might expand in the heat of the kiln, causing cracks and permanent swellings to appear in the fired piece.

Air bubbles in plastic clay interfere with the fingers' pressure, particularly in throwing. Unwedged clay has uneven consistency—there are harder and softer areas. The harder or dryer parts dry and shrink sooner, the softer or more moist sections lose their water more slowly. This condition may result in warping and cracking during firing. Uneven consistency also disturbs the action of fingers shaping the pot. Some commercially prepared clay is so well done in the factory that it requires little or no wedging in the studio.

A wedging board or table has a work surface made of very thick plaster of Paris or unpainted wood—sometimes covered with sturdy canvas. Clay does not stick to these porous surfaces. A diagonal cutting wire is attached. (Construction of this handy equipment is shown in Chapter 17.)

The potter stands facing the wedging board.

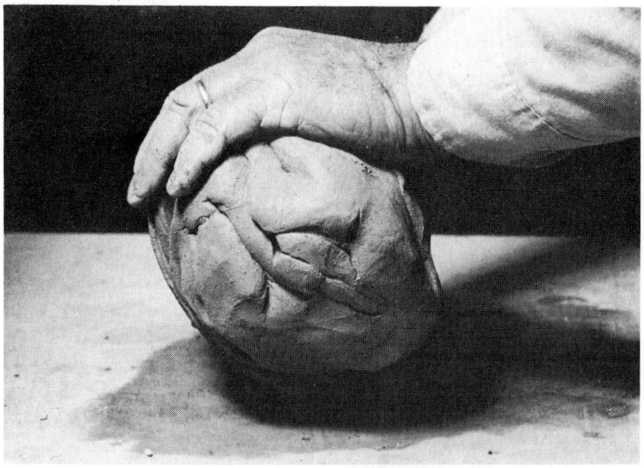

5-5–5-7

WEDGING BY KNEADING

Kneading the clay around itself into a spiral mass is one way to wedge clay.

5-5

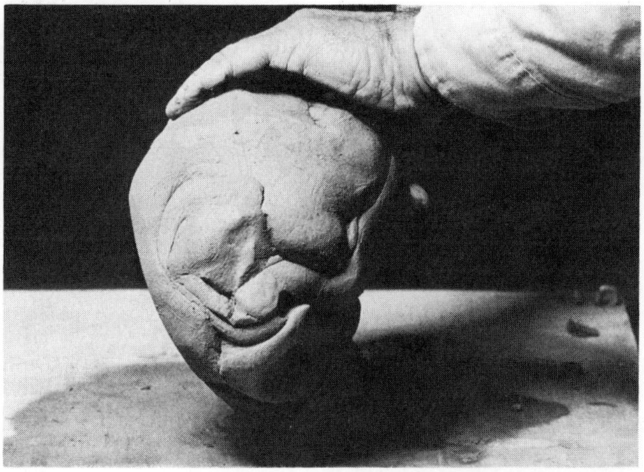

5-6

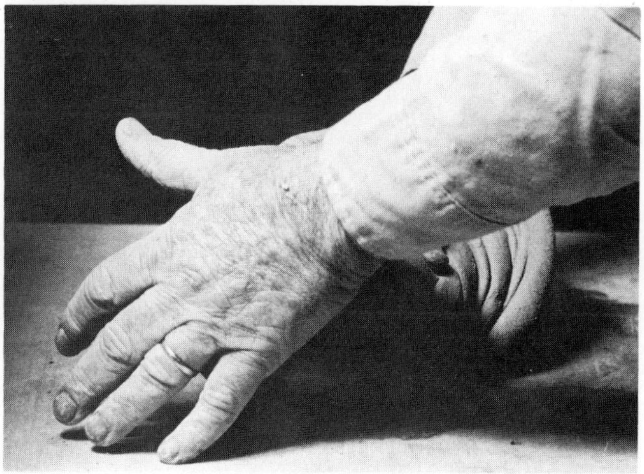

5-7

5-8–5-13 *WEDGING ON A WEDGING BOARD*

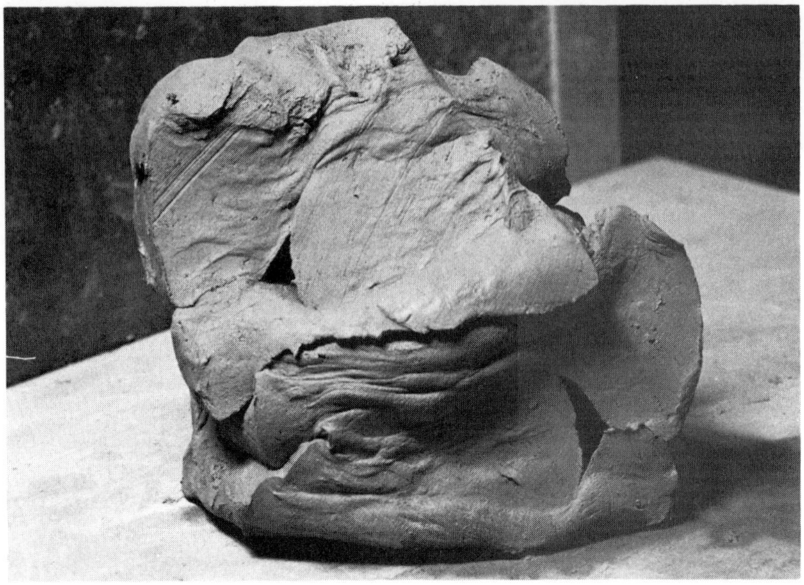

5-8 *The clay before it is wedged.*

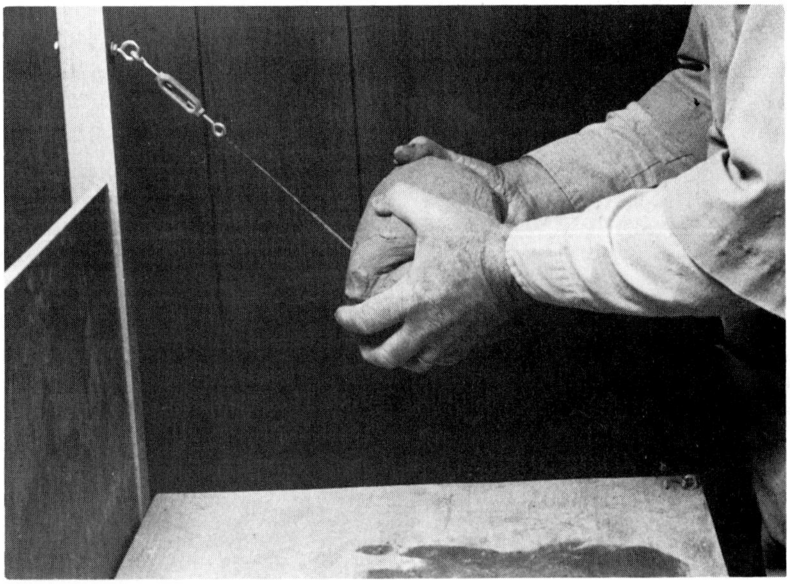

5-9–5-10 *Cutting clay across a wedging board wire.*

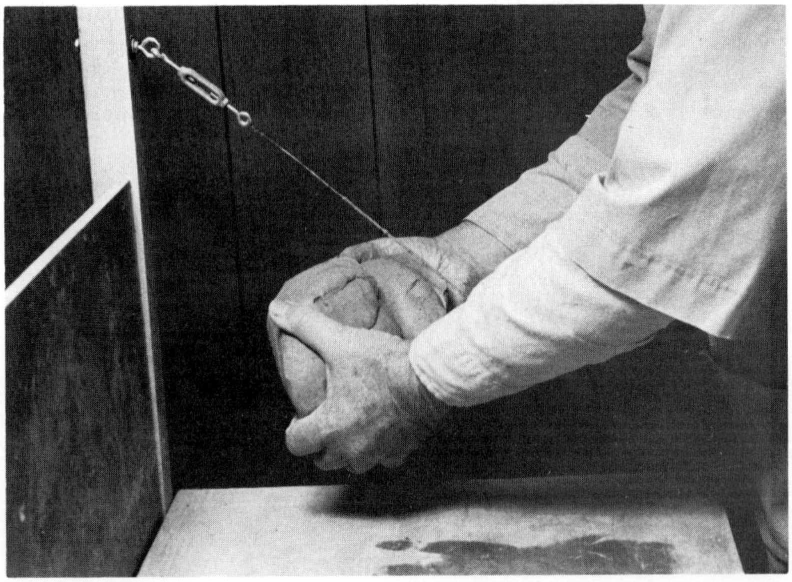

5-10

There are two favored wedging techniques, and many potters combine the two rather than use one exclusively.

One method is to knead the clay firmly on the table, turning the mass around itself in a squeezing, rolling motion, again, much as if kneading bread dough (figures 5-5–5-7). Apply increasing pressure each time and with the same hand. Observe how the clay churns around itself in a slight spiral or snail-like shape. Some potters consider this method of wedging particularly well suited to the preparation of clay for throwing, as it results in the clay platelets arranging themselves in spiral alignment, consistent with the alignment of platelets produced by circular pressure on the potter's wheel.

A second wedging technique requires the use of a wedging table with a cutting wire attached (figures 5-8–5-13). With both hands, hold as large a ball of clay as can be managed. Bring the ball down through the cutting wire, to slice it in halves. Throw one of the halves down on the wedging surface, the cut face exposed to one side. Now throw the second piece hard on top of the first, its cut surface also exposed to one side. Repeat the cutting and the hard slamming, but each time cutting at right angles to the previous one.

In this way every possible clay surface is exposed and the air bubbles are broken and expelled by the thrusts of the masses against each other. Hard and soft parts are repeatedly cut and mixed. In a short time the uniform clay, free of air pockets, is ready. Pulling a finger tip gently across one fresh face reveals no inequalities in texture, no disturbing air pockets.

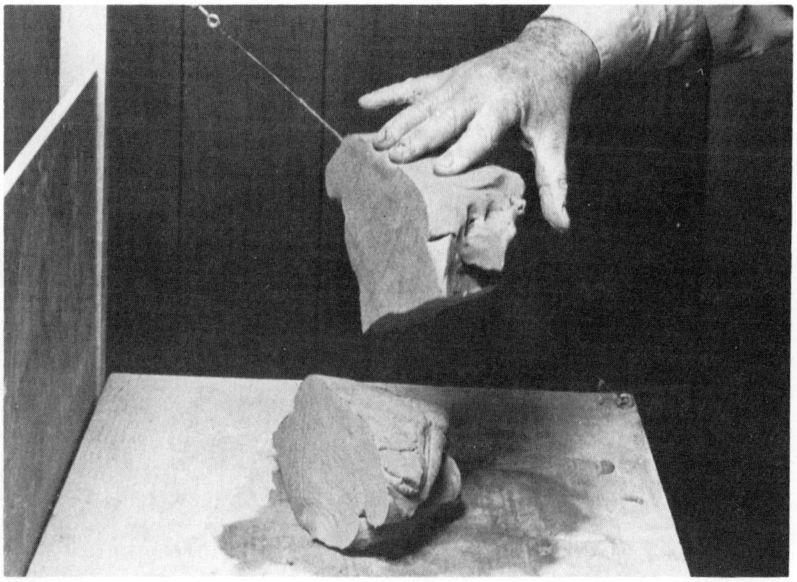

5-11–5-12 *The second piece is thrown hard on top of the first piece.*

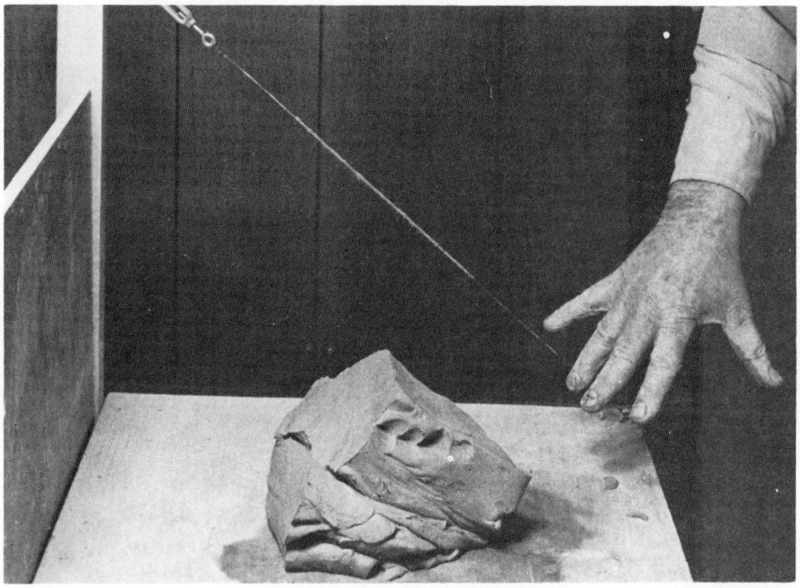

5-12

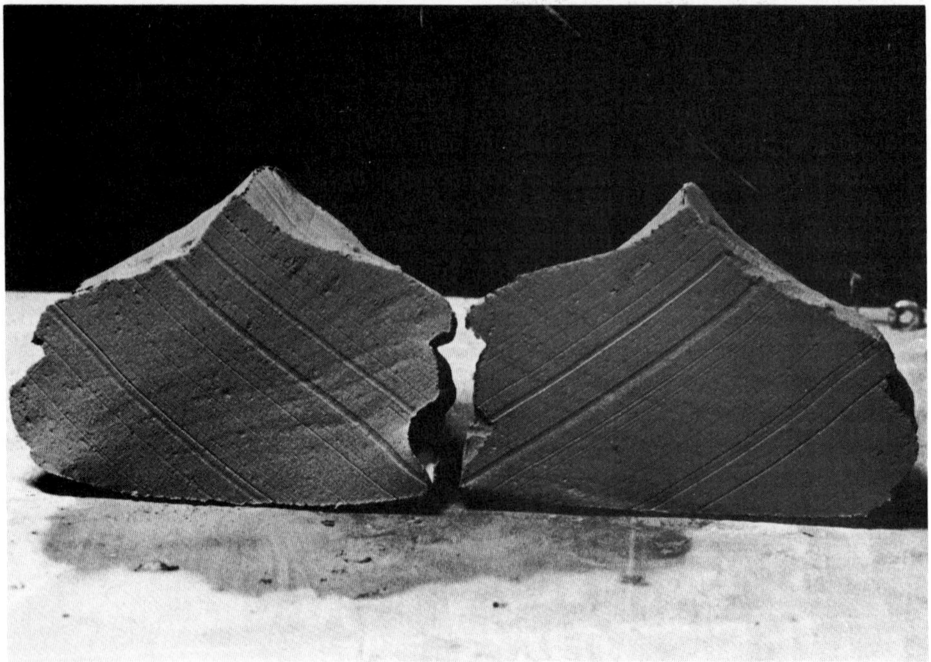

5-13 *Wedged clay.*

6 THE PRACTICE OF HANDBUILDING

CHARACTERISTICS OF HANDBUILT POTTERY

Nearly all the potter's creative and technical work is done by hand. Nevertheless, the term *handbuilding* refers uniquely to certain techniques for shaping clay, mostly by hand, without the use of a potter's wheel or other machine. In throwing on the potter's wheel, finger pressure on rotating clay gives the pot its distinctive personality. The machine gives direction to how the potter's hands will work and thus forces its own mark on the pot. But in handbuilding we expect those traits that come from finger pinching, rolling, flattening, pounding, scraping, carving, smearing, squeezing, and other hand and arm movements applied directly by the potter to the clay form. A few simple tools may be used, such as a potter's knife, sponge, modelling tools, scrapers, paddles, and a round stick or rolling pin. Other tools may be improvised. In recent years a variety of mechanical and electrical machines for making slabs and extruding coils has appeared on the market; they are mostly employed in busy potteries and large studios.

In some of the more traditional folk-based pottery cultures, the strictures and conventions established over a long history remain strong to this day. They have been passed down from generation to generation and are rigidly adhered to. But for many modern potters, each handbuilding method furnishes an opportunity for innovation and creative flexibility. The technical rules to be observed and the skills to be practiced are few; the opportunities for invention are infinite. For the most part, attention may be confidently focused on the developing pottery form and its surface.

The few rules or skills have mainly to do with the condition of moisture in the clay. For example, when fastening two pieces of clay together, both pieces should be softer than leather hard; otherwise they will not remain fastened through drying and firing. If the clay is too dry and stiff, then the pot resists easy shaping. The walls of a piece sag if the clay is too moist. Therefore, the percentage of water in the plastic clay added or removed during clay preparation, wedging, and working the clay is critical. Clay that seems a bit stiff can be made softer by sprinkling some water over it while kneading on a slightly moistened work surface. Keep the pot from drying by covering it with dampened cloths or plastic bags.

Compared to throwing, handbuilding generally permits and even encourages leisurely contemplation of the work, judgements and responses, revisions and changes as it evolves. There is ample opportunity to walk away from the piece, leaving it to rest for a while (covered so that it will not dry), then return with a fresh view and keen, new insights.

Although the following methods to shape clay by hand are described separately, it is not only possible but very often more interesting to combine techniques. The interplay of individual personalities can be quite exciting.

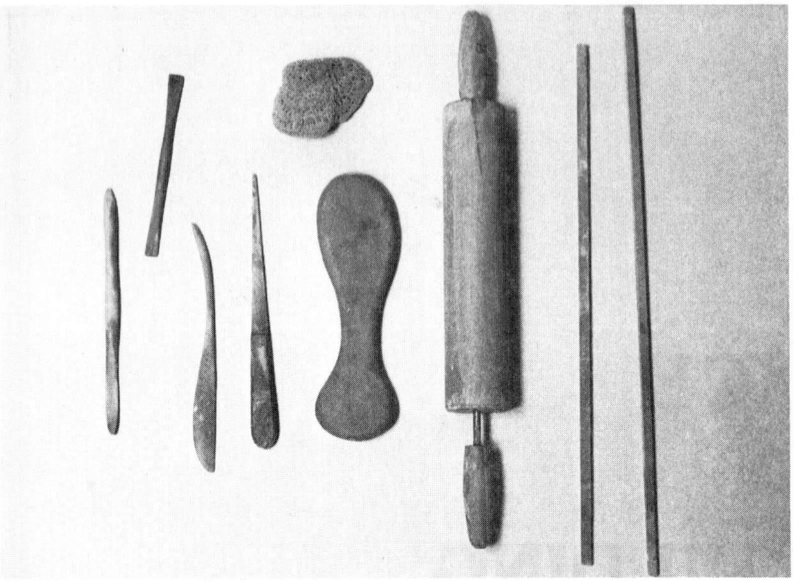

6-1 *The tools for handbuilding are few.*

6-3 *Bailey Mini-Might Slab Roller.*

Bailey Pottery Equipment Corporation, Kingston, NY.

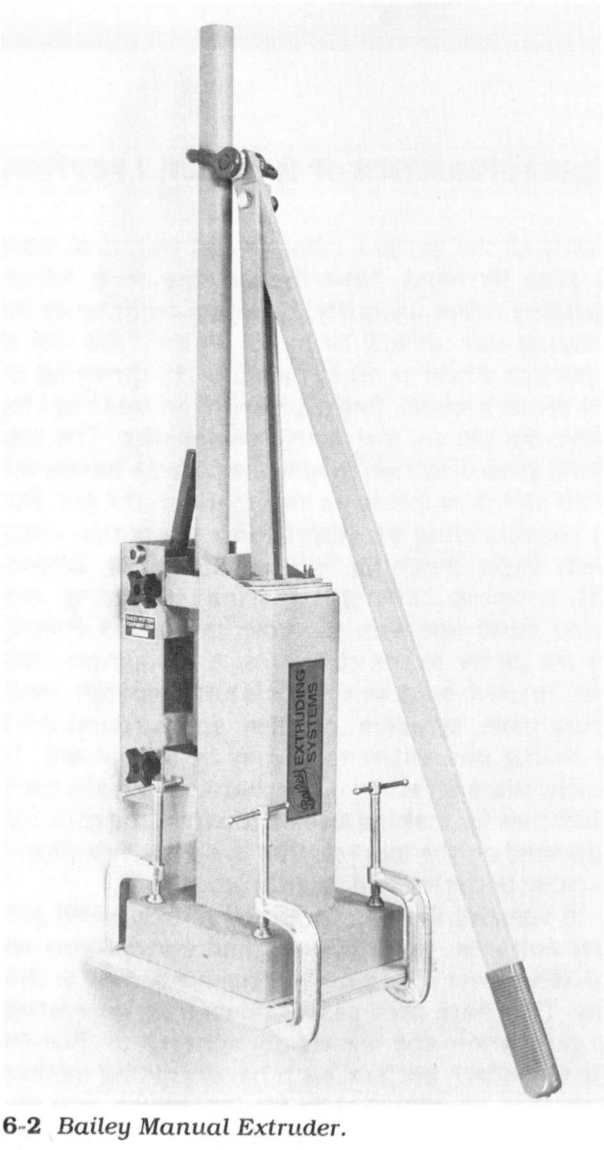

6-2 *Bailey Manual Extruder.*

Bailey Pottery Equipment Corporation, Kingston, NY.

PINCHING

If the reader has had little or no experience in working with clay, pinching is an excellent introduction to the sense of plastic form. Many potters remember pinching pots as their very first exercise. Pinching may very well be the oldest method man has used to shape clay, for it is the most straight-forward contact the fingers can have with the plastic material. If the hand grasps a wad of clay and the fingers squeeze tight shut, the squeezed shape released may very well assume the shape of a fetish, a figure, a god, or in contemporary terms, an abstract sculptural form.

Cup a baseball-size ball of clay in one hand, push the thumb of the other hand into it, and then begin pinching around the hole, slowly rotating the ball with each pinch. The series of pinches and the rotation gradually bring into being a bowl, a gourd, a shell, a deep form—volumes cradled in the supporting cup of the potter's hand.

Pinch precisely but gently; do not crack the wall. Pinch until the wall is thin enough; pinch until the wall is thick enough. Wall thickness is right when the weight of the pot, hefted in the hand, presents no surprises. This is a useful method to determine proper wall thickness in all forming techniques.

Use small balls of clay for small pots and larger pieces for bigger pots. All the clay goes into the pot. The shaped volumes can be joined side to side, a cluster of grapes or a bunch of bananas. Two shapes can be joined rim to rim to produce an entirely different form. Small pinch pots can become supporting bases and tops for other pots. When at-

6-4–6-7 *PREPARING A PINCH POT*

6-4 *Cup a ball of clay in one hand.*

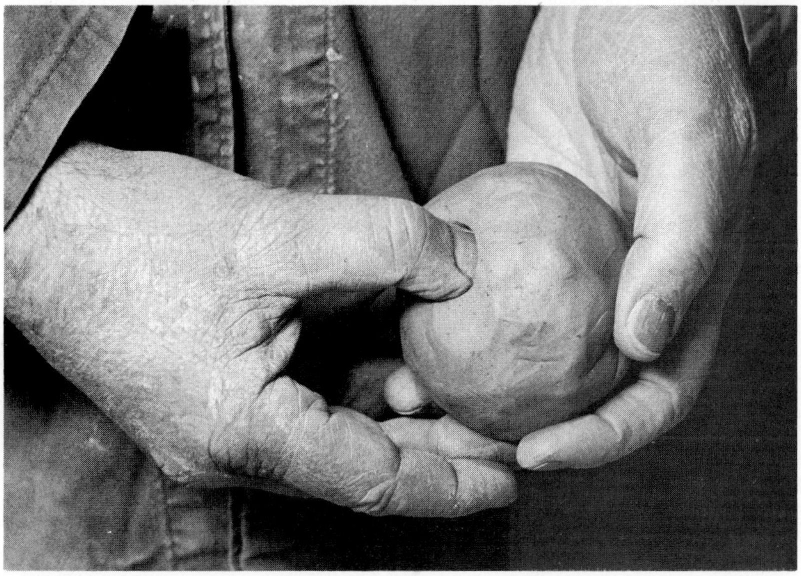

6-5 *Push your thumb into the clay ball to make a hole.*

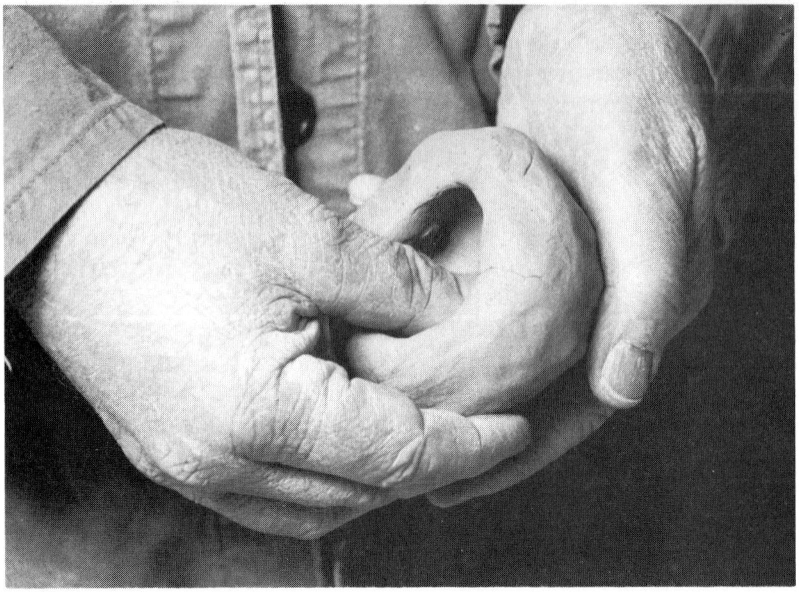

6-6 *Pinch the clay wall, turning the ball with each pinch.*

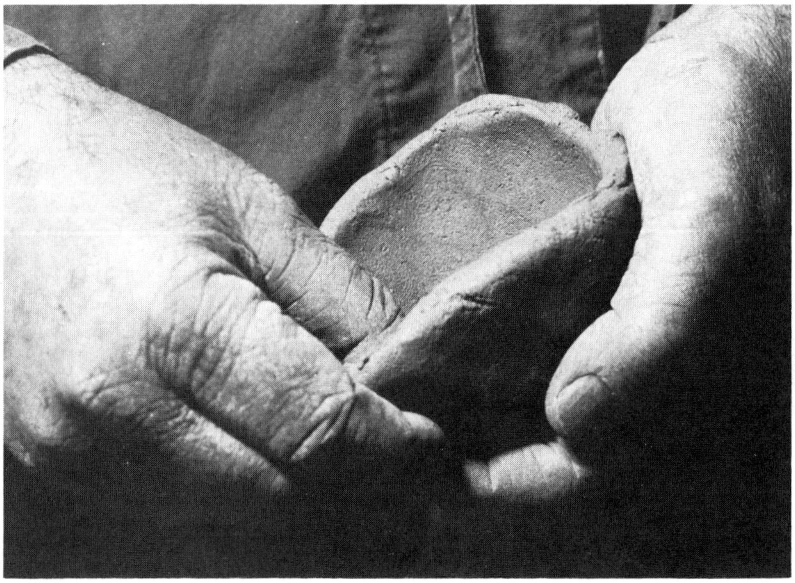

6-7 *Pinch precisely but gently to form the pot.*

6-8 *A good, uniform wall thickness with rounded rim.*

6-9 *Two varieties of pinch pot.*

6-10 *A vase assembled from four pinch pots.*

taching two forms, scratch the touching surfaces and rub them with a wet finger until some of the clay rises to the surface as a thick slip. Then firmly squirm the pieces together at the slipped surfaces. Suddenly the clay grabs, and the union becomes firm; the separate pieces have become one. The experienced potter recognizes this universally applied joining technique by the term *score and slip.*

6-11 *A string-wrapped paddle.*

For a bigger pot, jam your fist into a larger ball of clay, then pat and pinch with the other hand around the clay enclosing the fist. It may be more effective to do the patting with a wooden mixing spoon, a flat stick, or a paddle. Wrap a string or burlap around the paddle, use a roughened stick or a small branch with the bark intact, and observe how the clay surface becomes varied and enriched. Perhaps the finger marks left by pinching can be integrated with the paddled textures as a complement to the form. It is so easy to do too much, to overwork the textures. The plasticity must remain fresh, not tired.

The vessel's edge is called a *rim* or *lip*. It should be smoothed and shaped by the fingers and a wooden modelling tool to give grace and definition to the form, much in the same way lips define the mouth. Study the pot to determine its posture. Should it sit straight up, or off to one side, perhaps leaning somewhat? Establish the base by setting the still plastic clay down on the work table with a little smack for flatness.

At their best, these pinched forms look as if they were found in nature. Unselfconsciously, they have grown out of plastic clay understood well by supporting hands and with pressure sensitively applied. In good pinch pots visual elements seem to follow the sense of touch, the tactile experience incarnate in the clay.

COIL BUILDING

Coil building is perhaps one level higher in sophistication than shaping clay by pinching alone. The coils or ropes are usually placed on a clay base, then on top of each other to make the pot tall. The separate rings are joined by score and slip. The walls may be refined and made uniform by smearing the coils together and pinching them into a consistent and even wall thickness and shape, accompanied by scraping. The pot's wall then becomes more consistent in thickness and can be made thinner. The same amount of clay goes further for a larger piece. The shape of the pot can be more precisely controlled.

The piece grows in complexity, sophistication, and fluidity, and each element becomes stronger as a visual statement. Perhaps coined really as terms of endearment for working with clay, human nomenclature is used for the parts of a pot: foot, belly, shoulder, neck, and lip. As in the human form, these structures are distinct in good pots but they simultaneously build upon each other to form an integral whole, with each part of the pot's anatomy complementing rather than overpowering; the foot providing a strong undergirding, the belly volume, the neck an upward thrust away from gravity's pull, and the lip giving character and definition.

Some potters subscribe to a more intrinsic concept of pottery form, in which the clay is considered to be a skin, surrounding the inner volume with tension. In this sense, the inner volume of air enveloped by the clay is really the form, a balloon of sorts. The inner volume is the major dynamic or force. The clay defines the form; it is the shell or surrounding skin which makes it possible for us to see the pot.

These two aesthetic models are among the many influences on potters' work. Other sources of inspiration are bodies of thought: cultural, sociological, political, financial, scatological, historical, and even nonsensical. The pottery student will recognize in many works an eclectic composition, while in others a single concept seems to dominate.

The following is a basic approach to good coil

6-12 *The foot, belly, shoulder, neck, and lip.*

building, which can be applied equally well to pottery and clay sculpture. The reader, however, should feel free to experiment and innovate.

A coil piece is usually worked intermittently, an advantage of coil building. Therefore, the building is more conveniently done on a small, portable platform. This makes it easy to move the piece from place to place in the studio without disturbing the clay. The small work platform is called a *bat.* Bats are sometimes made by pouring plaster of Paris into a mold such as a pie tin, or it can be a square of wood, masonite, or some other strong, porous material. Since clay would stick to a non-porous surface, probably resulting in cracking and warping during drying and shrinkage, metal, plastic, and floor tiles make poor bats.

Shape the base of the pot by hammering with the side of the hand a small ball of clay, which has been placed in the center of the bat. Pound gently with the hand until a flat pancake of clay about 1/2″ thick develops. Later, the base will be partially hollowed out from underneath to make a footring. For now, trim the base with a potter's knife to the desired outline, round or free form as might be required by the pot's shape. If sculpture is being built, the base may be omitted; just starting with the first coil may be all that is needed.

From your clay supply, tear off a chunk and squeeze it between the two hands into the shape of a fat sausage. Lay the sausage down on a clean, slightly damp, unpainted wooden table—this is the ideal clay work surface—and roll with the outstretched fingers, not the palms, back and forth to make a long coil. Spread the fingers and keep them straight. Simultaneous with the rolling, move the hands slowly apart towards the opposite ends of the lengthening coil. Every few rolls, flip the coil around a bit for the cross-section to be more or less round. Perfectly round coils are unnecessary; the effort would be a waste of time. Many coil builders consider a fat coil, 3/4″ or so in diameter, to be better suited to fast, good coil building. Nevertheless, some prefer to use pencil-thin coils and make fine pots.

Score and slip the periphery of the base where the first coil will be attached. Firmly squirm the coil into position into the slip around the base. Cut the two loose ends for them to join at the correct length, and attach the two ends to each other by means of score and slip. Roll another coil. Score and slip the top edge of the first coil, and squirm the second one on top of it. After the second coil is in place and joined to the first coil, rub and smooth clay down from second to first and first to base, over the exposed joints all the way around inside and outside.

Squeeze and pinch the wall gently and precisely between finger tips to develop the precise silhouette, wall thickness, and shape. Paddles may be used in the same manner as with pinch pots. Firm but sensitive handling is required, as great vigor cracks and damages the clay.

Add a third coil, then a fourth, and continue until the pot's mature height is reached. With each new coil score and slip, then pinch, paddle, and scrape the new coil into conformity with the refined, completed wall below. When a new coil is put in place, the shape at that level must be given its exact form. It will be very difficult to return to it and attempt corrections.

Between work periods, the piece should be well covered with plastic bags or plastic film to keep it from drying. However, should the clay happen to dry beyond leather hard it is still possible to return it to a good working consistency by wrapping it with damp cloths or slightly wet newspapers. Gradually, the clay absorbs some of the water from the wrapper and plasticity slowly returns. The process of rewetting is unhurried; it may take a day or two or longer for good sized pieces. If the cloths are too wet, the clay absorbs too much water, and the pot will collapse.

Pinching and paddling are excellent ways to re-

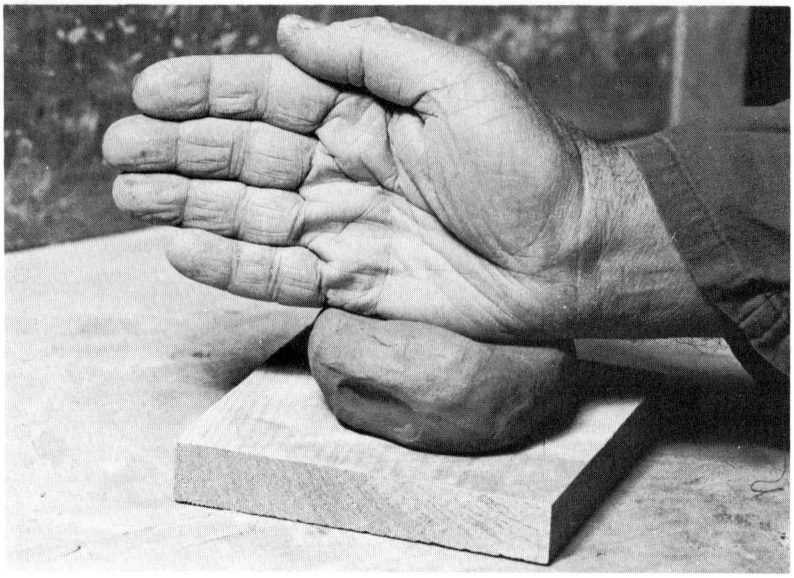

6-13–6-17 *PREPARING A COIL POT*

6-13 *Shape the base of the coil pot.*

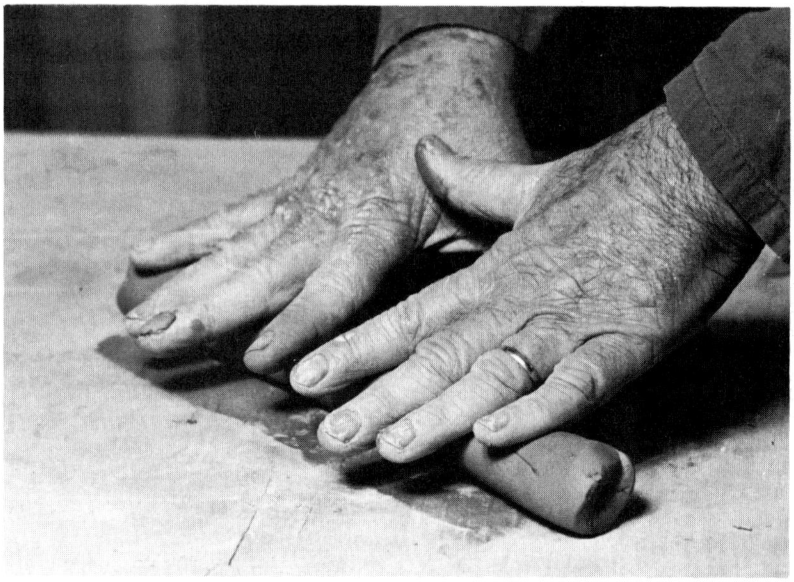

6-14 *Roll a coil.*

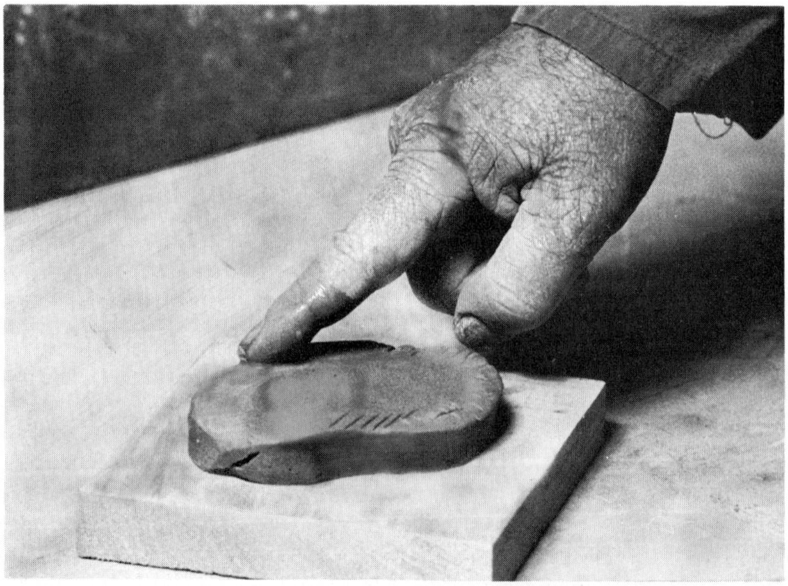

6-15 *Score and slip the periphery of the base.*

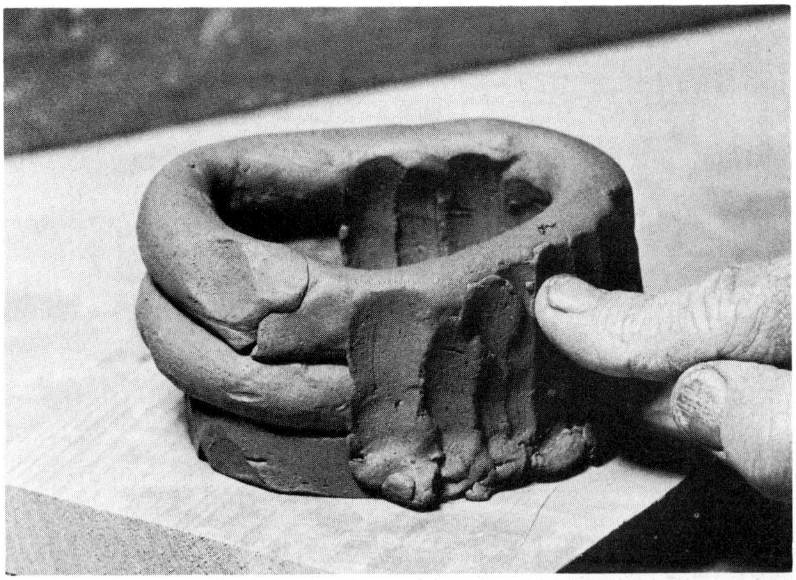

6-16 *Start the wall with the first two coils.*

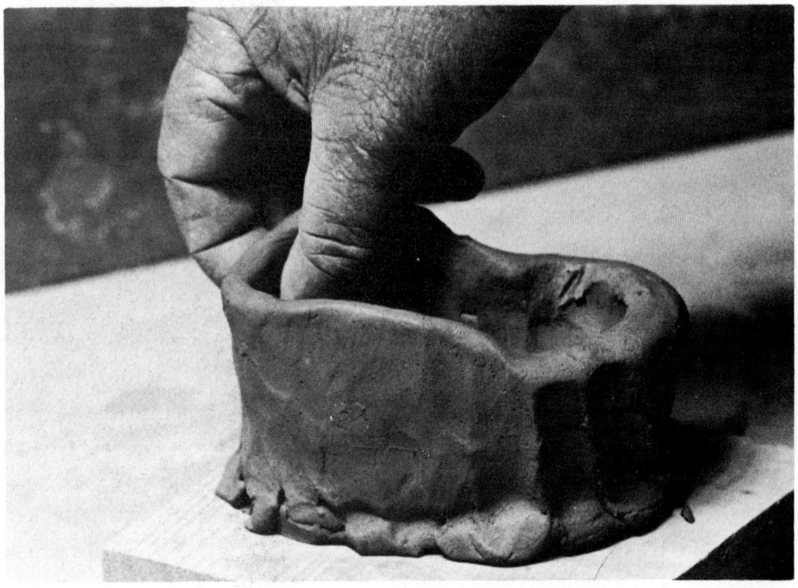

6-17 *Pinch the wall to the proper thickness and shape.*

fine the form of a coil pot. As a by-product, they give to the piece characteristic surfaces: from pinching sensitively and uniformly executed a patina of finger marks finely distributed in a lovely rhythm, from paddling a network of overlapped, textured facets. At their best, these surfaces are revealed as a natural setting to the form. On many pots no other surface enhancement is needed. The way the pot was made, honestly displayed, seems for these pieces a right and fitting resolution.

However, other textures may be more deliberately applied to the surface and made to work well with form. Scraping the wall produces a more refined contour and smoother surface, best done when the clay is dry beyond plastic. Use any tool that will do the job: a knife edge, modelling tool, or spoon edge. There are thin metal scrapers made specially for the purpose.

Burnishing imparts a polished, mirror-like surface, most successfully accomplished when the clay is drier than leather hard. Smooth, polished pebbles and spoon backs are excellent burnishing tools. Rub one small part of the pot's surface at a time in a tight, circular motion, gradually extending the burnished area. Burnishing actually compresses the finer clay particles and pushes the coarser particles below the surface. The result is a polished texture, and in the fired state the clay is slightly more water tight. Good examples of burnishing may be seen in the traditional pottery of the Southwestern Pueblo Indians. Scraped, smooth, or burnished surfaces probably look best on forms which are precise, with extremely refined silhouettes.

Texture is the tactile quality of surface. Clay is remarkable in its inexhaustible capacity to reveal sensitively an infinite variety of textures, be they

6-18 *An asymmetrical coil pot.*

6-19 *Pinched and paddled textures.*

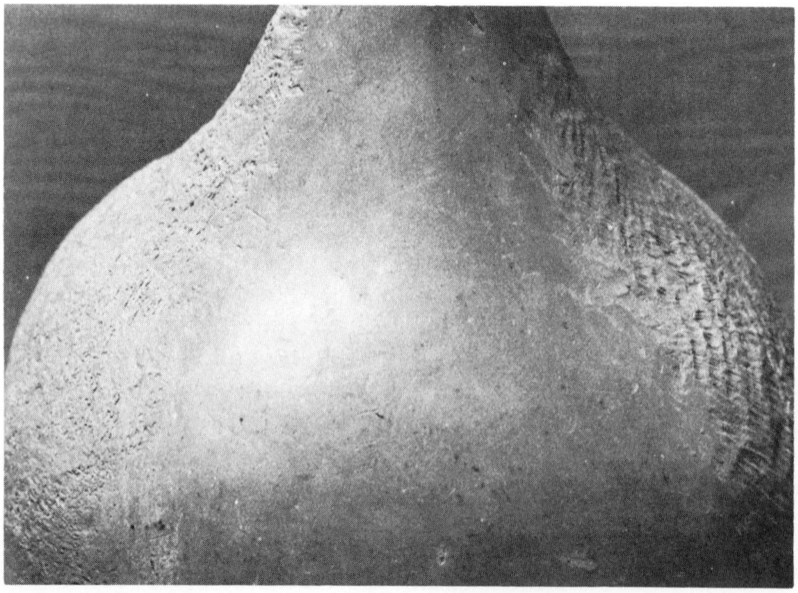

6-20 *A burnished surface.*

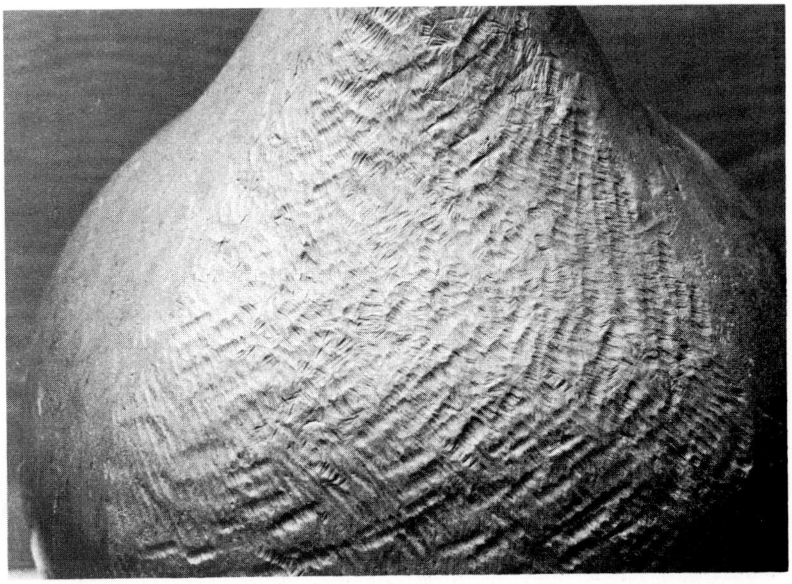

6-21 *The texture of tightly twisted cord wrapped around the paddle.*

intrinsic or applied. While working on a piece, there is sometimes concern as to whether additional textural decoration would be suitable. An often-used criterion to decide appropriateness is whether the added textural material seems a natural outcropping of the form—is its appearance reasonable? Potters are sometimes carried away by cleverness and skill. Virtuosity may assume a meaning separate from the pot, and the piece may lose its vitality.

There will come a moment of recognition while coil building when the form and its size happily coincide. This is the unmistakable signal that there has been enough coiling, pinching, paddling, and scraping. The form is in precise alignment between bottom and top, yet base and rim are incomplete. Feet and rims are the beginnings and the endings in pottery. The foot is the pot's foundation; its definition may be found in the rim. The rim unites the inner volume of the piece with its exterior content. It is the pivot for the eye's movement.

By now, after having worked on the piece for some time, the clay's consistency is rather firm but should not be drier than leather hard. This is the ideal condition for working on rim and foot, the rim first. Consider the rim as more than merely the end of the wall: It is the point at which the inside volume meets the outside surface. It is the crucial edge that the eye traverses as it digests the whole pot. It is a pivot point. We expect the rim of a symmetrical piece to conform in its own symmetry, although there are exceptions. Conversely, free-form or asymmetrical work—pot or sculpture—is usually capped by a free-flowing edge.

Shape the firm clay of the edge by scraping, pinching, and smoothing with the fingers, wooden

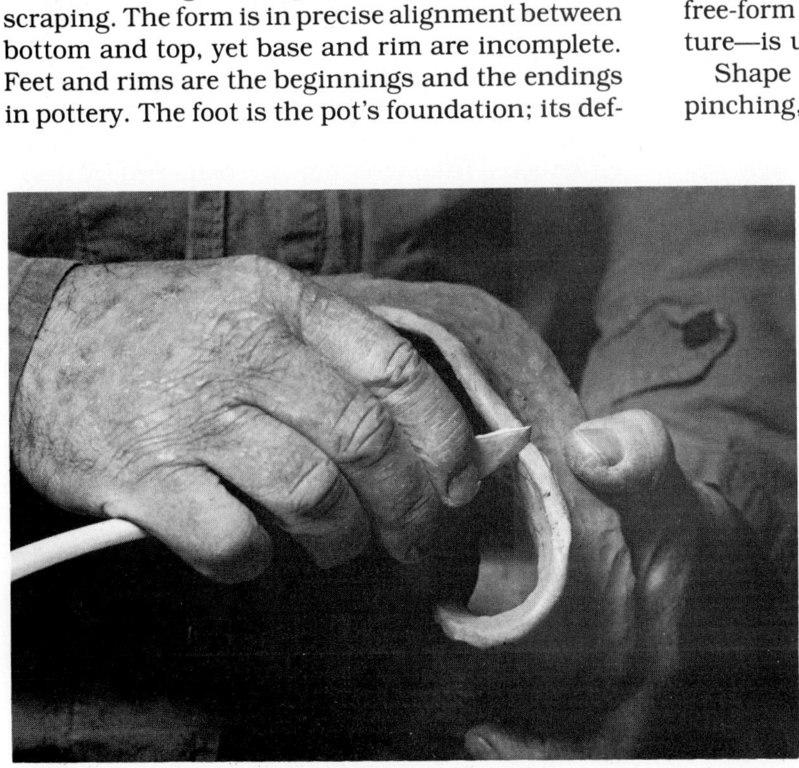

6-22 *Shaping the rim.*

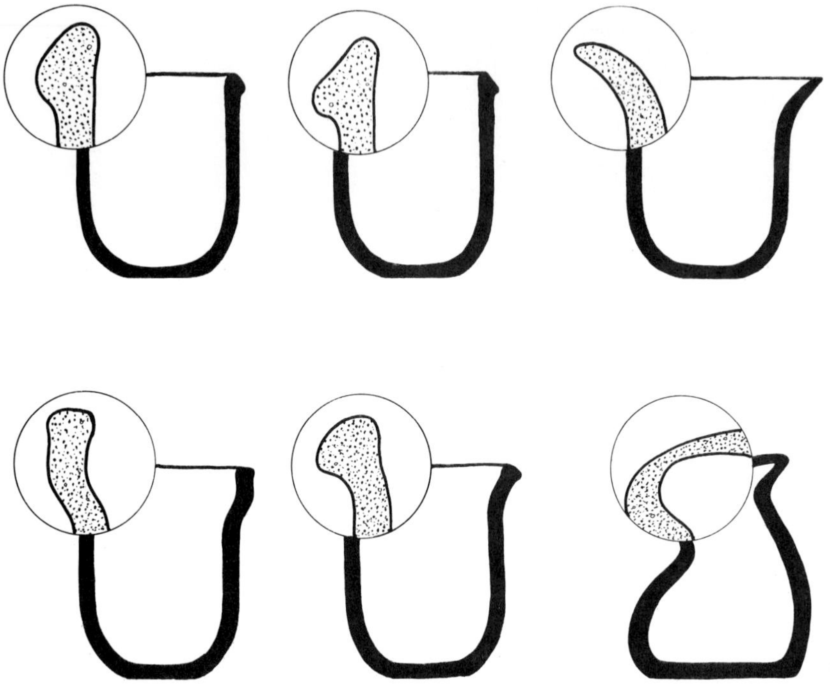

6-23 *Some rim variations.*

modelling tools, or a knife. If the edge is too thick, pinch it thinner and scrape clay away; and clay may be added to even out a thinner section. Every once in a while, run your finger lightly around the rim as you work on it, feeling for minute lumps and bumps, interruptions that might otherwise escape the eye. Avoid sharp edges, for glaze does not stick to them well, and they tend to chip easily. Slightly thicker and rounded in cross-section, as in a tin cup, a beaded edge resists chipping, holds the glaze nicely, and above all furnishes a distinctive visual accent at the top of the pot. A shadow will be created just below the slightly thicker rim, which sets the pot off as in a border or picture frame. However, some pieces seem to benefit from a rim that is thin, delicate, and precise in delineation. Most symmetrical rims are absolutely true and uniform in thickness. A free-form rim may have parts and sections thickened to accentuate its irregular rhythm. With the fingers and modelling tool add and work in extra clay where the rim's fluid motion requires thickening, until the top of the piece has been shaped and refined completely.

The vessel's base is so important to its visual structure we call it the foot. It supports the body substance. There are two parts to the foot: (1) the silhouette viewed from the side—the very bottom of the pot's wall and (2) the surface underneath the base, seen only when the piece is turned over. In some pots the base silhouette is elevated as if on a pedestal. There can be many variations of pedestal footings. Another category of footing is a swift, linear flow upwards from the bottom directly into the wall; the pot's silhouette is a continuous sweep of curves.

A good footing gives the pot an identity separate from its environment but not a self-conscious isolation. In this way the pot is perceived to have a life of its own apart from the world. Some pots want a strong uplift, a more vigorous surge and mobility. Other pottery forms are comfortably matched with solid, secure, stable footings. At their best, parts of pots are in a state of essential compatibility with each other, the rims to the bodies to the feet. Sometimes there is the suggestion of tension, or even a slight contradiction within the framework of the composition. Other pots seem to be in a state of grace in their inner adjustments. Therefore, thought about the details of the foot should involve its relationship to the whole form.

Turn the coil pot over on a work table, resting on its rim. The piece may be cushioned by a soft pad of cloth or foam. Some potters prefer to hold the pot cradled securely in their laps. A wire-loop end tool, wooden modelling tool, and perhaps a potter's knife are needed. Draw a line around the bottom to outline where the base will be. With the modelling tool, scrape and cut away all the extra clay from around the outline, until it is precisely defined. At the same time the wall an inch or so above the base should be carefully shaped to reveal the proper silhouette. Scrape and cut a little at a time. A symmetrical form usually requires the same kind of symmetry in the base, whereas a sculpture or free-form vessel generally has an irregular, eccentric base conforming to the total composition.

In most utilitarian pottery, the base is hollowed out half way through its thickness, to create a footring. This is the reason for having made the base 1/2″ thick. Hollowing out the inside of the base

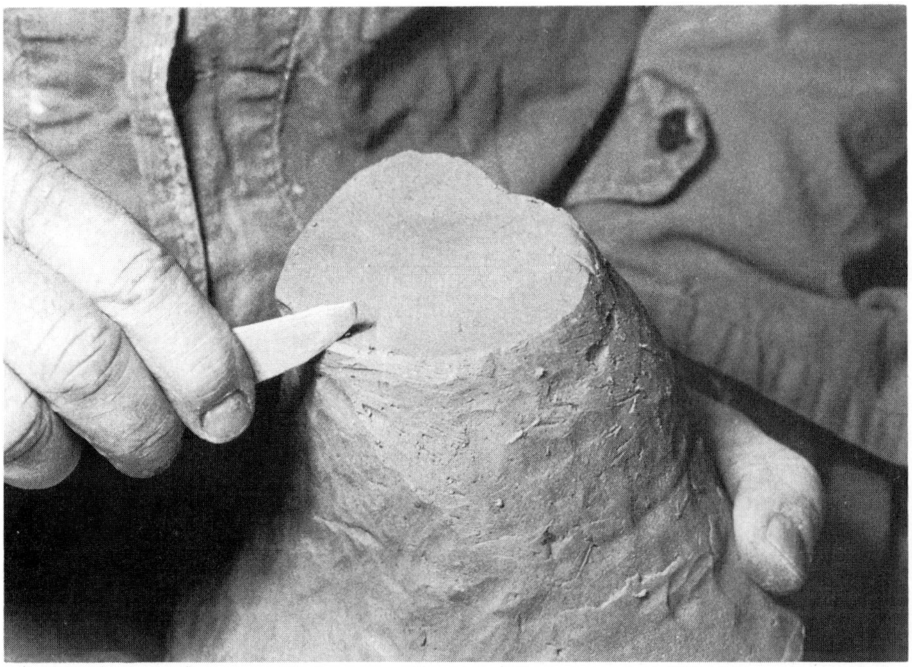

6-24 *Scraping and cutting clay to shape the footring.*

6-25 *Hollowing out the base inside the footring.*

leaves a nice, neat footring circumnavigating the periphery, and the pot actually sits on this ring. The hollowed out base and footring, particularly in pieces of very wide diameter, furnish a cross-section which resists warping and permits almost the entire base, except for footring, to be glazed.

An alternative to the hollowed out base is to make it just slightly concave. Press the base in from underneath with your thumb, or, if this is not practical, scrape some of the bottom away to secure a sound footing. For this type of base, concave but without a footring, the base thickness should be the same as that of the walls.

A pot's footing is not as visible as the rim. The underside cannot be seen except when the pot is turned over. Nevertheless, a good pot is distinguished by consistent, careful craftsmanship in all parts. For the potter and the pottery connoisseur, delight in excellent workmanship contributes immeasurably to the overall quality of life in the piece.

6-26 *A hollowed-out base and footring.*

6-27 *A concave base, without footring.*

SLAB BUILDING

Pinching pots is intimate, coil building is slow and introspective; in slab building forms appear suddenly out of the assembly of separate sheets of clay.

A variety of slab rolling machines, operated by hand and electricity, are available to the potter. These machines quickly produce slabs of almost any thickness and area. Small slabs can easily be made by rolling the clay with a round stick or a rolling pin. It is convenient, but not essential, to have two sticks as thickness guides, placed on each side of the clay; one can then work the rolling pin over the clay down to the level of the sticks. Slabs may also be formed by pounding clay with a mallet or your hand, or by slicing away at a block of clay, using a taut cutting wire.

There is no accepted standard for the thickness of slabs, but about 1/4″ may be a good figure to

6-28–6-38 *CONSTRUCTING A SLAB POT*

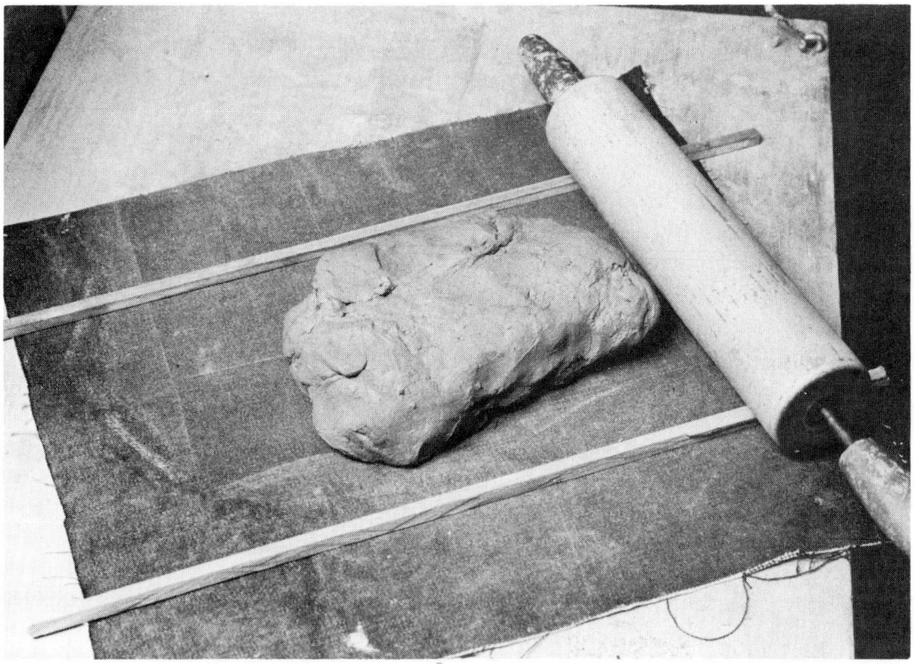

6-28 *A rolling pin, thickness sticks, and a sturdy cloth.*

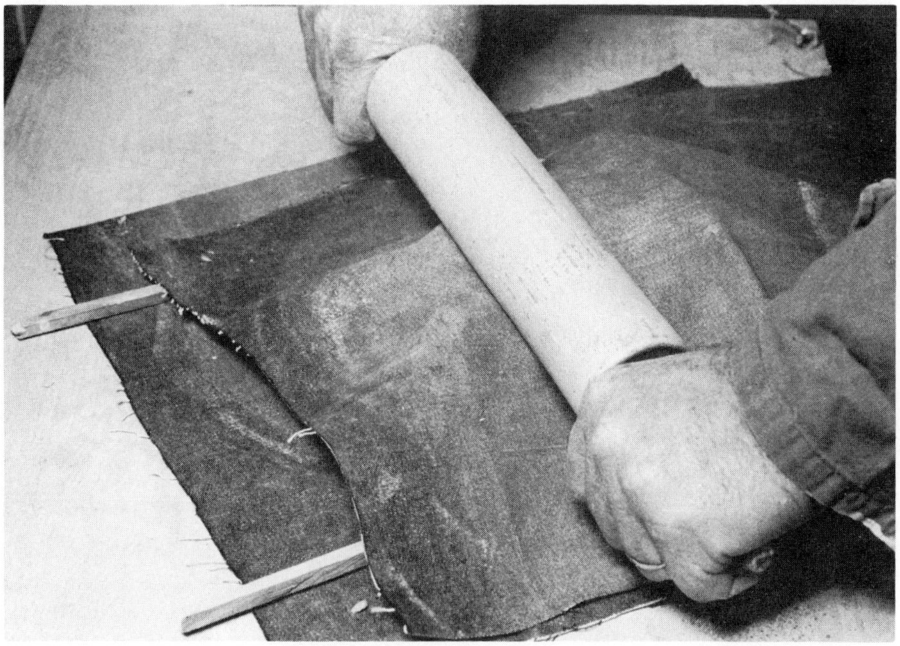

6-29 *Rolling a slab.*

start with. Larger, more massive pieces or coarse grained pottery bodies may require thicker slabs, while smaller, more delicate constructions or fine grained bodies may use thinner slabs. When rolling slabs, a layer of burlap, canvas, muslin, or a sheet of newspaper placed below and over the rolled clay keeps it from sticking to the table and the rolling pin.

Some slab pots are constructed when the slabs are still quite soft and plastic. They can be easily peeled away from the cloth or newspaper layer. Other types of slab forms are put together from slabs which have been permitted to become firm, almost leather hard. After being rolled, the slabs are slowly dried, perhaps even under a light cover. They may be left on the work table or moved to a safe location for the drying.

The simplest slab form is made from a single sheet, taken up while still soft and brought around itself by hand, creating a new volume. Perhaps the

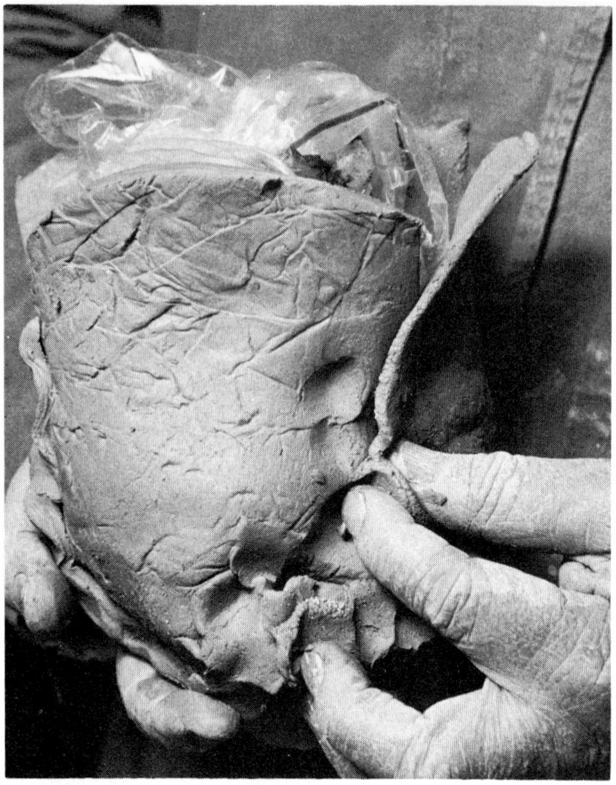

6-30 *Soft slabs wrapped around a core of crumpled plastic film.*

6-31 *Slab pot set aside to dry.*

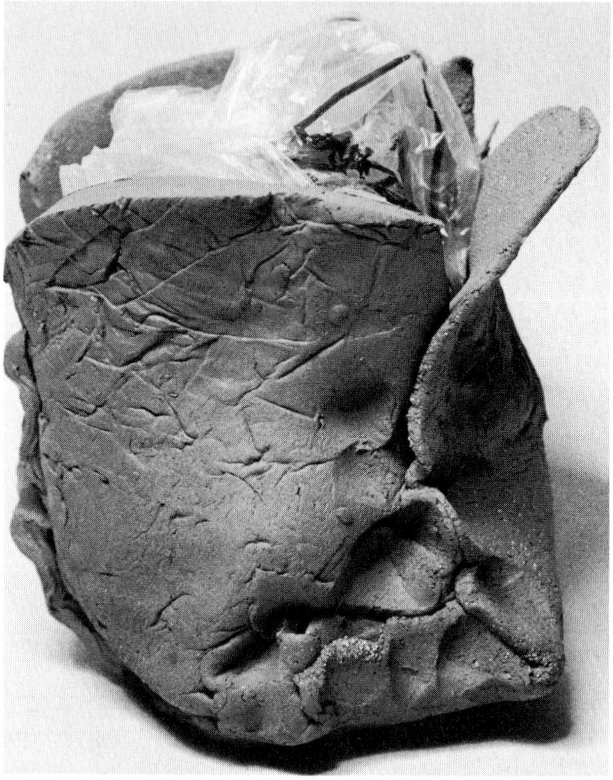

6-32 *Slab pot on a slab base.*

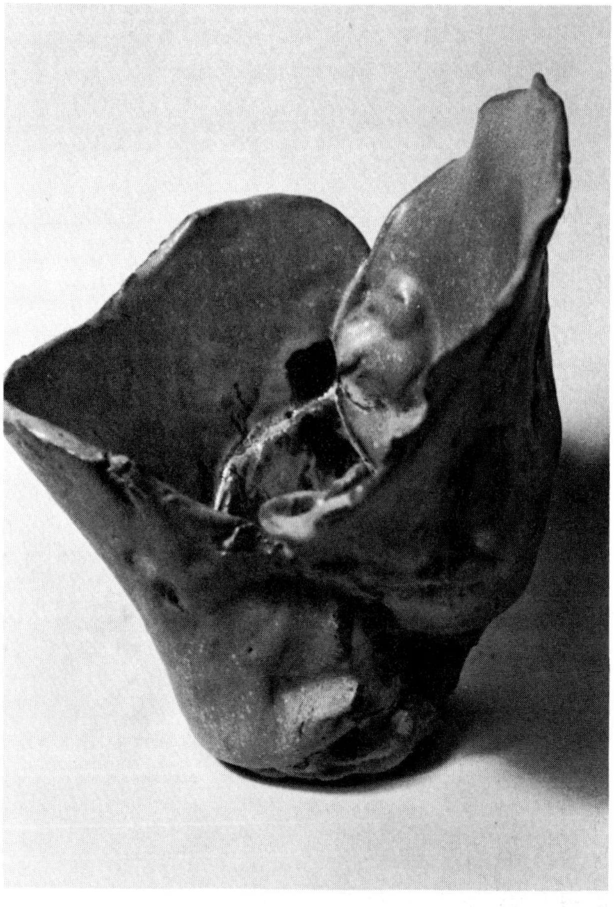

6-33 *Slab pot after the glaze firing.*

6-34 *A canteen shape. The neck is a slab, the body is from pressed slabs, and the base is a pinch pot.*

slab might be wrapped around a ball of crumpled newspaper, a balloon, some soft foam, a tube as from paper towels or a pipe, or some other improvised inner core. If one slab is inadequate for the form, then use two or three or more, overlapping them a bit, and fastening by score and slip. The inner core (armature) must have some give to it, to accommodate shrinkage around it of the drying clay. Otherwise, the pot might crack or come apart. Combustible armatures may be left within the slab construction, for the material will burn away in the kiln. However, if the core is non-combustible, it must be removed before the pot dries and shrinks.

Another method involves arranging slabs in the hollow of a mold form, perhaps a wash basin, or in the hollow of a hammock of cloth suspended between two chairs. The plastic slab may be laid over a convex form—a boulder, a watermelon, a gourd, a rock, or even someone's shoulder. The slab dries and assumes that shape. Use it alone, as it comes from its mold; or several of the slumped and pressed parts can be assembled by score and slip to make a completely new structure. For example, a canteen shape assembled from two shallow bowl pieces can be fastened together edge to edge, then stood on end with slab rings at the top and the bottom for foot and rim. For an interesting visual interchange, include coil and thrown parts in the composition. Some potters fabricate plaster or styrofoam molds for pressing and slumping slab building. Characteristically, these soft slab assemblies have the look of plasticity, a tentativeness that is the mark of the hand, and a kind of softness of spirit.

More meticulous and exacting in the mechanics involved, slab building can also be executed by joining firm, almost leather-hard sheets. It is prudent to plan ahead very carefully. A good procedure is to prepare a preliminary paper or cardboard pattern, as in a dress pattern. Lay the pieces of the pattern over slabs, and cut the slabs to shape with a potter's knife. The slabs should be permitted to dry till firm, but not quite leather hard. Were the slabs assembled while soft, the pot would slump and sag; but if too dry, the joints would come apart. Press the slabs firmly together at their joints, using score and slip. If possible, paddle the joined corners with a flat stick for even more pressure to secure the joints. It is a good idea to fasten along the inside corners a thin coil, a fillet, which is then smoothed to fill and reinforce the corner.

Made as it is of flat components, this slab construction has straight sides and defined corners, rather than soft, curvilinear volumes. The forms can be rectangular, trapezoidal, or irregular and box-like. Characteristically, there is an architectural cleanliness and precision about this kind of structure, much as if plywood were being used rather than clay. Some potters have developed techniques of slab construction which incorporate fiberglass and other fillers in the clay, an innovation to permit building unusually large pieces.

A work made of slabs can hardly be mistaken for any other style of pottery. Assembled from sep-

6-35 *A cardboard pattern for a slab pot.*

6-36 *Score and slip the joining surfaces.*

6-37 *Press the slabs firmly together.*

6-38 *Secure the joints wherever possible by inserting a clay fillet.*

6-39 *Textures rolled into a slab.*

arate sheets of clay, the forms are extended planes and surfaces. Those large surface areas give a prime opportunity for working with texture since each surface is merely an empty facade, much like a billboard. It tempts the potter to leave his mark. Before rolling the slab, some texture-producing materials may be spread out under or on top of the clay slab. The rolling pin pushes them into the clay surface. Straggled string, paper doilies, crumpled newspaper, bits of dried clay, rice grains, sawdust, wood shavings, dried weeds, grasses, and leaves are but a few of the found materials that spring to mind. Mix and rearrange them to create an endless variety of patterns and textures. Placed sensitively, their major impact is to enhance the three-dimensional form. A temptation is to do too much, to develop a surface that is too busy, profuse, and too strong, an urge the sensitive potter resists.

Rims and feet in slab constructions should be carefully considered and executed. These parts may be modelled and carved directly into the slabs or separate pieces may be attached. Necks, rims, and feet may be built of slabs consistent with the construction of the pot, but they may also be coiled and thrown. Some very interesting pottery and sculpture result from mixing of techniques all in one form: soft slab parts, hardened slab sheets, pinched and coiled sections, and thrown parts.

6-40 *Identical slab forms with rim and feet variations.*

CARVING AWAY

Pottery and clay sculpture forms are developed in one of two ways: by either building up or removing clay as the work takes shape. Coil building and slab building are good examples of the first. In these the shape grows as a result of coils and sheets winding, wrapping, and folding around the potter's perception of space. In these assembled pots the visual aspects of volume are frequently the major dynamic, with volume giving character and meaning to all other visual elements. However, when the piece is developed by carving and removing clay from a solid, other visual qualities become evident.

Put together a mass of clay of approximately the same proportions as, but slightly larger than, the desired final form. It might be a cube, a rectangular box, a pyramid, a column, an egg, a sphere, or any other solid form. After wedging the clay thoroughly, stand at the work table and thump the clay on the table until the mass takes the desired shape. It is more convenient to have your work on a bat. The tool techniques are quite simple: merely cut, scrape, gouge, and hack the clay away. Probably most useful is the wire-end modelling tool, but knives, spoons, wooden tools, and even fingers are good as well. It is prudent first to prepare a small model or clay sketch as the guide for the larger work. It will give a general direction of how to develop the shape from a clay mass and helps avoid errors. However, it is better not to try to copy the model exactly, for changes and alterations will be suggested by the work in progress.

For vertical round forms, such as columns or totem poles on which linear elements writhe and squiggle on the surface, leave intact an inner core. Cheese-box forms scooped out can become giant

6-41 *Form carved from a block of clay.*

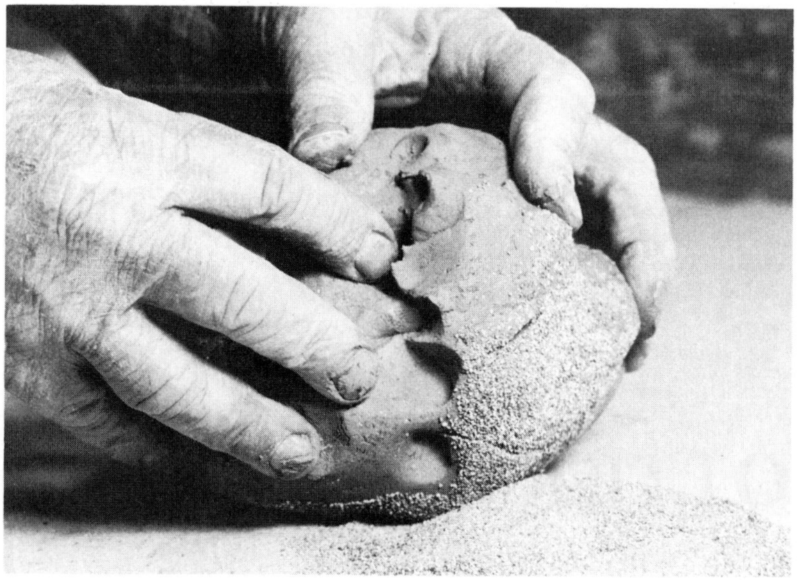

6-42 *Combine grog with clay by kneading.*

walnut shells; cubes carved away become massive boulders. The final piece takes its cue unmistakably from the shape of the original large mass. Often we perceive the outline of the original bulk mysteriously present, even after the clay has been removed. An aura of solidity is retained. Surfaces and textures in well-made pieces faithfully reflect how the tools shaped the surfaces, leaving lovely patterns of marks made by gouging, cutting, scraping, and scooping.

Wall thickness in these carved vessels is usually quite varied. Because of the method of shaping, the uniform wall thickness so characteristic and desired for coil and slab building need not exist in these pieces, and wall sections may be quite thick. Therefore, the clay composition should be open and porous to permit easy drying without excessive shrinkage and cracking. Some natural clays and pottery bodies are suitable because of their coarse grained formulation.

However, for more predictable control and safety in drying and firing, potters frequently mix with their clay a large quantity of grog. Since grog has already experienced its total shrinkage, it produces an open pore structure in the clay to which it is added. Although it also decreases plasticity, this is no disadvantage in this forming method. Grog imparts a strong, gritty texture. The mesh size of the grog is chosen on the basis of whether a coarse, medium, or fine gritty texture suits the work. Since plasticity is not a primary consideration for this forming technique, a large amount of grog can safely be incorporated, as much as one third of the total. It is easy to knead grog into the clay. Sprinkle some grog on the work table and over the clay mass. Churn and knead the clay with the grog until it is thoroughly mixed. Continue adding more grog until the desired quantity has been introduced.

For a different appearance, incorporate instead any one of a number of combustible materials, such as sawdust or coffee grounds. They also open up the pore structure for good drying, but their major contribution is to enrich the texture of the clay. They burn away in the kiln during bisk firing, leaving distributed in the clay mass and on the surface a conglomerate of holes the shapes and sizes of the added particles. The end result is somewhat like a dry sponge. The pot is made lighter in weight. It is usually a very exciting contribution to the effect of the piece.

7

THROWING ON THE POTTER'S WHEEL

Some potters find throwing pots on the potter's wheel the most gratifying way to work with clay. The reward is both personal and artistic: the laying on of hands around the soft, wet clay as it is revolving on the wheel, then seeing the form grow, nourished by your own person. The appeal is different from handbuilding. The speed with which things materialize, the considerable practice and formidable discipline essential to success, and the wheel's circular movement, limiting forms to round shapes, evoke other responses.

Although the potter's wheel was first used about 5500 years ago in Egypt, there are still a handful of pottery-making cultures, including some in the Western hemisphere, where the potter's wheel is not a part of the pottery-making tradition, even though the wheel does have other functions in the culture.

POTTER'S WHEELS

Readers of this book will probably encounter three types of wheels: (1) electric-powered wheels, (2) kick wheels, and (3) treadle wheels.

The *electric wheel* is driven by an electric motor whose speed is controlled by a foot or knee mechanism. The potter generally sits at the wheel facing the work. The wheel rotates counter-clockwise, a custom in this country and in some other parts of the world, that may be traced to historical factors in the early development of the potter's wheel.

The *kick wheel* is constructed of a heavy flywheel at the bottom, connected to the wheel head at the top by a shaft or axle, held fast by bearings to a sturdy frame. The potter sits at the wheel slightly above the work and puts the wheel into centrifugal motion by thrusting a foot at the flywheel. Most potters shape the pot with their hands while their feet rest, then alternately kick the wheel back to a good speed while their hands are at rest. The heavy flywheel rotates for a period long enough for the hands to work.

The wheel head of the *treadle wheel* is connected to the flywheel by means of an eccentric camshaft. Thus, a foot lever attached to the shaft is pushed back and forth by foot to rotate the wheel head. There are some types of treadle wheels at which the potter stands, while others provide for a seat facing the work. The treadle wheel is kept in constant motion, but at a slower speed than required by kick wheels.

There are some qualities one looks for in any potter's wheel, no matter which type may be selected. The frame or case should be substantial in weight and strength, to avoid vibration and looseness of parts. The wheel design should preferably permit the potter to throw facing the work, sitting slightly above or standing, and there should be adjustments for throwers with long legs and throwers with short legs. The speed control should be variable to allow sensitive and immediate regulation of the wheel's rotation. Although a beginning pottery student does not at first use large amounts of clay, with progress sizable pieces will eventually be thrown. It is wise, therefore, to choose a wheel that can handle without strain at least 20 to 30 pounds of clay.

There are two methods to hold the clay on the

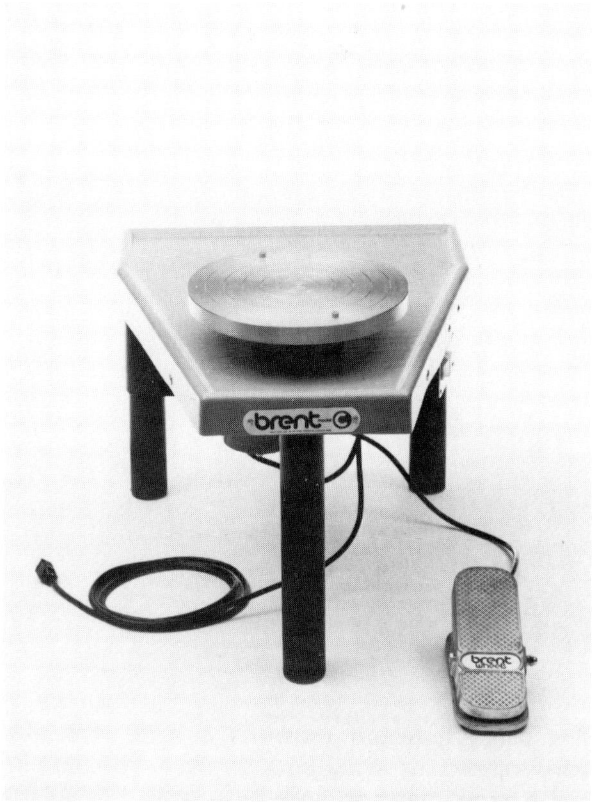

7-1 *Brent electric wheel, model C.*
Robert Brent Corporation, Indianapolis, IN.

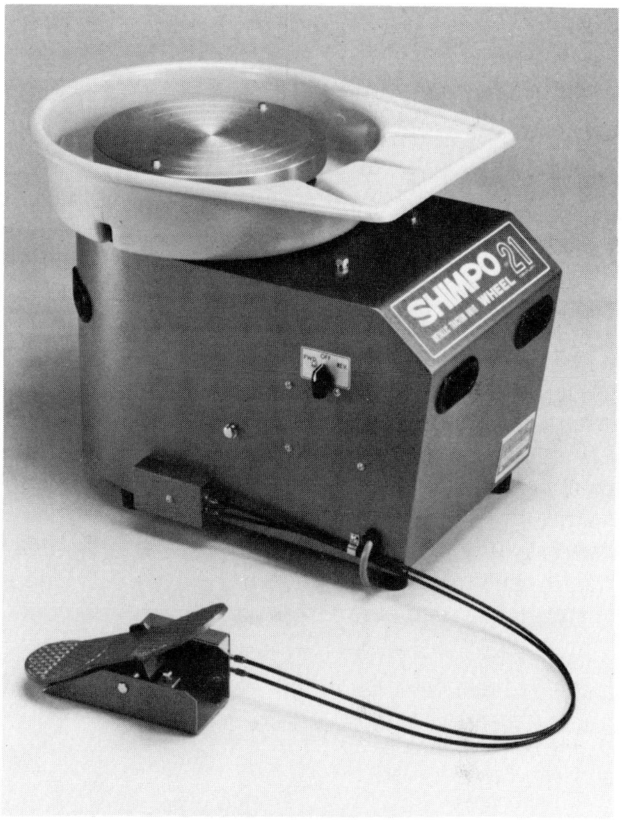

7-2 *Shimpo electric wheel, model 21.*
Shimpo-West, Inc., City of Industry, CA.

7-3 *Brent kick wheel, model J.*
Robert Brent Corporation, Indianapolis, IN.

7-4 ***Randall kick wheel, model H.***
Randall Pottery, Alfred, NY.

wheel head. On some wheels, the heads are flat discs. This arrangement permits the potter to remove a freshly thrown pot by cutting it through at the base with a taut wire or cord. The soft, wet pot is then carefully lifted by hand from the wheel head and set down elsewhere for drying. Some pots will be inadvertently mashed before this practice is perfected.

7-5 ***Treadle wheel no. 4.***
American Art Clay Co., Inc., Indianapolis, IN.

The second design requires a head specially shaped to hold throwing bats in place. The bats fit into the wheel head and are held snug. Throwing is done directly on a bat. When a pot is completed, the bat is conveniently lifted from the wheel head with the pot safely on it. The bat may be easily returned to the wheel head where it automatically centers itself. The bat system has many advantages, but the method chosen is really a matter of personal preference and working arrangements in a studio.

CENTERING

All the clay on the wheel goes into making the pot. Therefore, the novice potter may find it easier at the start to practice with smaller balls of clay. Have at hand a bowl of water, a soft sponge such as elephant ear or silk, wooden modelling tools, and a cut-off needle. A clay ball that fits nicely inside the cage made by the hands, loosely joined, is a good size to begin with. If a plaster bat is being used, make it medium wet. Moisten the bottom of the clay ball. Facing the wheel, get it going rapidly in a counterclockwise rotation and almost simultaneously throw the ball down onto the wheel head or bat hard to make it stick. It will stick, but, as the wheel rotates, it appears to veer and wobble. In the early stages of practice it may be reassuring to fasten ball to bat before the wheel is set in motion.

Centering is the first step in throwing. It is probably the most critical stage. By direct pressure applied between the hands the irregularly revolving ball is brought into true, concentric rotation, perfectly round, with no trace of wobble. Dribble water off your sponge to lubricate the turning clay. The revolving clay should always feel slick wet under the fingers. Anticipate the first signal of slight dryness or friction felt by the fingers and instantly apply more water. Add only enough for lubrication—do not drown the clay, but keep it wet.

On a kick wheel the speed of rotation is greatest during centering. Electric wheels and treadle wheels may be rotated more slowly at this stage.

Potters have different centering methods. One good way is to place the hands on each side of the revolving clay, the heels of the palms pressing in towards the center, the fingers cupping the ball shape, the thumbs stretched out flat across the top of the ball. Leverage is improved by having the one hand pulled slightly around to the front, the other hand moved just slightly around towards the back. Push firmly towards the center of the wheel, with the heels of the hands assisted by the wrapped around fingers. Simultaneously, press the thumbs flat down from above. Do not hook or bend them—that would make a hole in the clay.

Push in towards the center between both hands, at the same time press down from above with the flattened thumbs. This is the force that centers the clay: down from above and in from the sides. You compress the clay into a centered position. Keep your arms and hands steady, and finally the clay ball will find itself centered. It is centered when it does not wobble. You can tell by looking, or by holding the tip of a point against the rotating ball. A centered ball does not bump against the point.

The practice for centering may use up many balls of clay. The clay may be used over and over again, but it should always be wedged thoroughly each time. Never re-use the clay without wedging.

7-6–7-10 *CENTERING.*

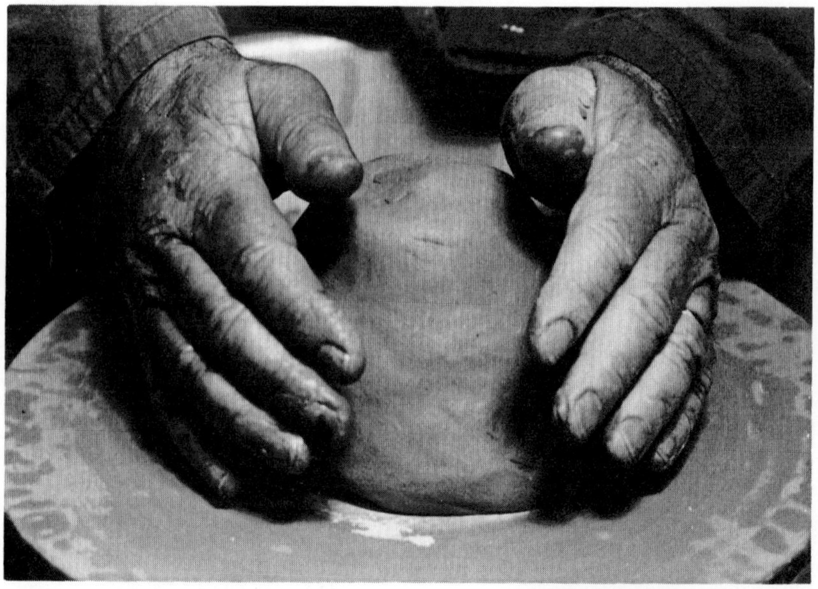

7-6 *Place the hands on both sides of the revolving clay.*

7-7 *Push firmly towards the center of the wheel.*

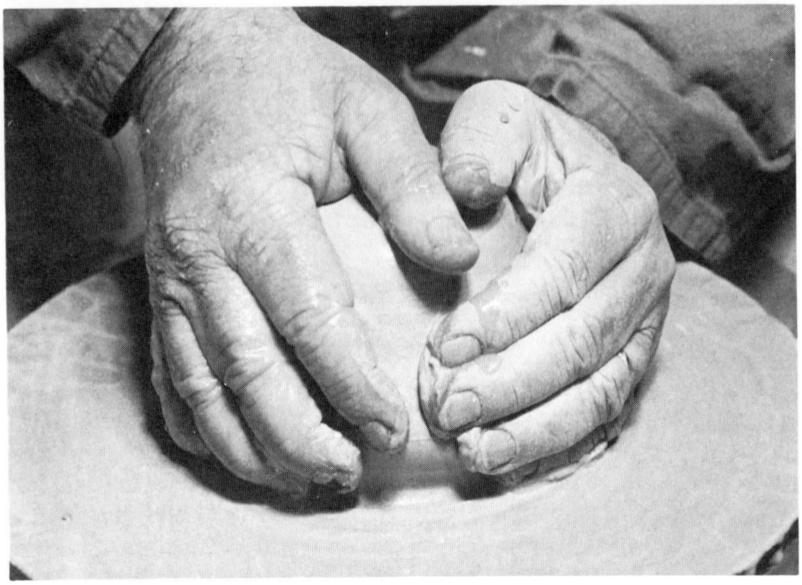

7-8 *Push in with both hands and down from above with the flattened thumbs.*

7-9 *Compress the clay into a centered position.*

7-10 *The clay is centered when it does not wobble.*

OPENING THE BALL

After successful centering, the next step is to open the ball. Every thrower has a favorite way to open the ball. One good method follows.

Rotate the wheel somewhat more slowly. With the left hand, cup the wet, centered ball, the left thumb sticking straight up. The three middle fingers of the right hand come down to the thumb and push it with steady, forceful pressure down into the center of the ball, making a hole all the way down to within one half inch of the bottom. This base thickness will be used later to carve a footring. Now make the hole bigger without making the ball wider by compressing the clay into vertical height. Your left palm presses firmly, to prevent the ball from spreading.

Practice holding the hands and fingers stable and steady, not wobbling and shifting. The piece must remain centered, and the hole in it concentric. The clay around the hole remains uniform in thickness all around. That is the difficult part. Remember constantly to refresh the water on the clay, to dribble water from your sponge or finger tips. When the centered ball has a hole opened in it equally centered, the work is ready for the next step, the raising of a cylinder.

7-11–7-13 *OPENING THE BALL.*

7-11 *With the three middle fingers of the right hand and the thumb of the left, push with steady pressure.*

7-12 *Make a hole to within one half inch of the bottom and widen it.*

7-13 *The piece must remain centered, and the hole in it concentric.*

RAISING A CYLINDER

Virtually every vessel form starts as a cylinder. It is the foundation from which other shapes may be developed. Narrow cylinders with tall, vertical walls can be transformed into pitchers and jars, whereas, wide, low cylinders become bowls and platters.

Connect the right hand to the left hand at the thumbs by locking the thumbs together with firm pressure or by twisting the thumbs around each other. Find a system to hold the thumbs together that is comfortable for your hands. Thumbs attached in this fashion act like a hinge, permitting the hands to pivot against each other. The finger tips can be brought together to touch, while the two thumbs remain locked together.

The wheel's rotation should be somewhat slower than the previous step. Place the left hand inside the hole in the opened ball and the right hand on the outside but on the right side. The wheel's rotation in a counterclockwise direction causes the clay wall to move fluidly through the space open between opposing finger tips. Keep the fingers of the left hand bunched together inside the hole and the fingers of the right hand bunched together outside the hole. Do not separate the fingers. Kept close together, the fingers brace and support each other. Bend the knuckles slightly (there are two knuckles in each of your fingers). The bending fingers bring finger tips into position on the clay wall, rather than hands and palms—all to the good.

One finger from the left hand inside pushes out against the revolving clay wall, one finger from the right hand outside presses in against the wall. Locate the pressure between finger tips opposite each other, on each side of the rotating wall. Slightly bent knuckles bring the finger tips against the clay.

7-14–7-16 *RAISING A CYLINDER.*

7-14 *The fingers of the left hand are bunched together inside the hole, and the fingers of the right hand outside. The knuckles are bent.*

7-15 *Locate the pressure of the finger tips opposite each other.*

7-16 *The finger tips come together, starting at the bottom and moving up in a slow, steady vertical climb.*

Use the finger tips to reduce friction and consequent pulling so that the clay glides easily. In this way, the throwing is more sensitive, fresher, and more precise. If the two opposing finger tips are squeezed harder, the wall becomes thinner and the vessel grows taller. Whenever the finger tips are removed from the work, move them smoothly and slowly away from the clay. The movement must not be abrupt.

Most often, the finger actually working on the clay is either your index finger or middle finger, one from each hand, not both. Opposing finger tips come together, each time starting at the bottom of the cylinder, and move slowly up in a steady, vertical climb as the clay revolves. They have assumed the function of eyes in that they sense the condition and thickness of the clay wall. This sense of touch is critical in throwing, and it will become more acute and accurate as your experience grows.

The inside finger tip starts at the inside corner between wall and base. The outside finger tip starts at the corner between bat and wall. The two tips, squeezing firmly, rise up the wall opposite each other. A complete pass is made, from bottom to top. Then make another pass and yet another, until the wall is uniform and the wall thickness appropriate for the cylinder's size. Inside and outside corners should be squared, not rounded or sloped. All the clay possible is brought up into the wall's height. Practice using the same amount of clay to make ever taller pots. While learning and practicing, cut the pot down in halves each time, using a cutting wire or knife to see the wall's cross-section. Thus, the sensitivity of the potter's finger tips will be verified.

The wheel's speed is adjusted to the movement of the fingers. When the wheel turns more slowly the fingers should move upward more slowly with

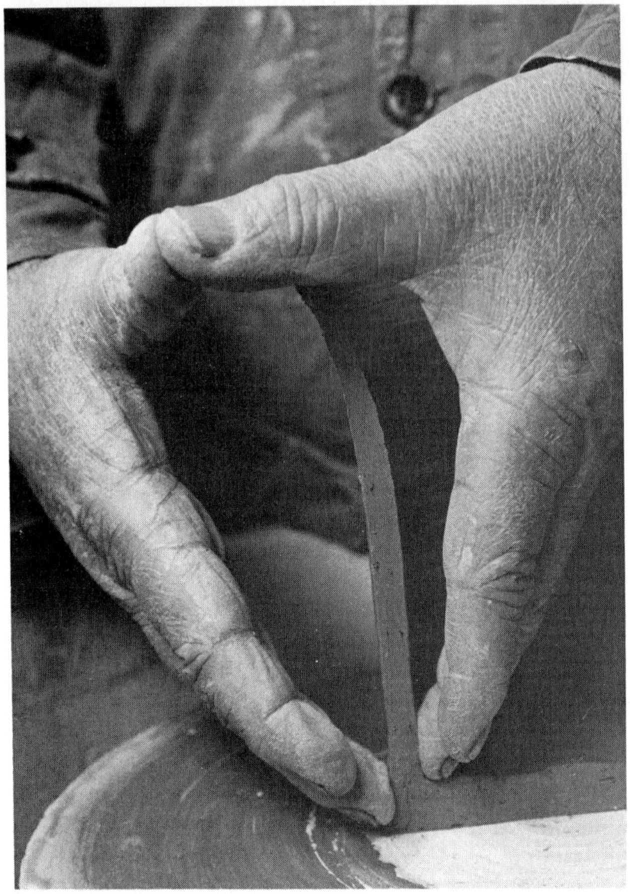

7-17 *Use this finger position to refine the cylinder.*

7-18 *Bring as much clay as possible up into the height of the wall.*

7-19 *A rib is used to refine surface and form.*

relaxed pressure. When the wheel rotates more rapidly the fingers may climb more quickly, pressing more firmly. It is vital that the potter learns effectively to utilize the direct and close relationship between wheel speed and finger pressure. It is a cause-and-effect relationship. The result is better control over form and texture.

Squeezing finger tips on revolving clay produce a spiral groove in the soft clay, called *throwing marks.* This textural pattern is a natural consequence of the throwing act. At their best, the spiral fingermarks rhyme with the form in accent and vigor. A perfect cylinder probably looks best when the fingering is repetitive, close, and uniform. Swelling and receding volumes, tight necklines, and sloping shoulders are each complemented by enriched outcroppings and surges of finger pressure. Throwing marks are visible as shadows defined by ridges; we see them because of their light and shade interplay. They can be tight and closely spaced, wide apart, shallow, deep, delicate, and faint. The potter soon learns how these important expressive elements in the pot's surface can be deliberately manipulated to reflect the sense of its form. Under a glaze, the well-thrown surface is a rhythmic element integrated with the life of the pot.

Occasionally, an unbroken surface may be more appropriate than the visible marks of throwing. Perhaps the potter wishes to strengthen the effect of a certain glaze or to furnish a suitable surface for some decorative technique. Some glazes and decorations work best and can be more expressive over smooth rather than ridged textures. An understated, gentle, and delicate form may be more expressive with an unbroken, smooth surface. Immediately following the final throwing, pass a wooden rib or rubber kidney over the surface of the wet clay as it is revolving on the wheel. The smoothing tool is held by the right hand, and while the wheel slowly turns the blade is lightly pressed over the clay surface. Throwing marks recede, and, if the smoothing continues, they become completely obliterated.

To make a handsome cylinder, well centered, taller than wide, walls uniform in thickness and appropriate for the mass of the piece, with square corners, and a pleasing rhythm of throwing marks is a significant achievement for the learning potter. However, the pot may not be considered complete until its foot and rim have been put in place. Rims and feet serve the same purposes in thrown wares as they do in handbuilt pots (Chapter 6). In throwing, however, they can be executed directly on the wheel.

RIMS OR LIPS

The pot's lip is fashioned while the clay is still plastic, in place centered on the wheel, and immediately after the form has been thrown. But first check to determine whether the top of the pot is level and true. Often, as a by-product of throwing, the rim may become uneven, ragged, and mushy with water. Before final shaping, the edge must be trimmed clean. The tool of choice is a cut-off needle, which may be an opened paper clip, a dissecting needle, a hat pin, a safety pin, or some other sharp pointed, needle-like implement. Hold the cut-off tool in the right hand, between the thumb and the first two fingers. With the pot rotating slowly, gradually pierce the wall just under the ragged part of the edge. The needle point goes through slowly as the clay revolves; the hand must hold the needle steadily. After two or three turns of the wheel, the needle point goes through all the way around. Now catch the

7-20–7-21 *TRIMMING A RAGGED RIM.*

7-20 ***While the pot slowly rotates, pierce the wall gradually just below the ragged part of the edge.***

7-21 *Deftly lift the trimmed ring of wet clay.*

7-22 *Finger tips shape a soft, rounded lip.*

7-23 *A chamois strip is used to smooth and round the lip.*

point on the pot's inside with the index finger of the left hand and deftly lift the trimmed ring of wet clay. The trimming is discarded. The rim is now sharp, clean, precise, and defined.

As the wheel turns slowly, three finger tips from each hand, clean and thoroughly wet, come down all together bunched around the trimmed rim. Close them together, as the six finger tips gently caress the rotating sharp edge into a soft, rounded lip.

Some potters prefer to shape the rim with a small strip of chamois leather, rather than with finger tips. The chamois strip is about one half inch wide and about two inches long. It should be thoroughly wet. As the wheel slowly revolves, the chamois is folded down around the edge, with the gentlest pressure possible. A smooth, rounded lip results. This method is particularly suited to edges for cups or mugs. The disadvantage of lips formed by chamois is that they are always the same, mechanical in their uniformity. Where more vital visual elements are sought, the lips are probably better formed with the finger tips.

A good rim might be shaped just a little thicker than the wall underneath, as in the bead of a tin cup. A shadow appears just below the rim's thickness to give it definition and strength. Other pots may require lips which are less emphatic, thinner, and more delicate. The lip is so essential to the well being of the form, that a sensitive potter studies the pot's nature carefully to create the appropriate shape of rim.

FEET AND FOOTRINGS

There are potters who, for philosophical or production reasons, prefer not to have trimmed footrings in their work. They may consider the time spent uneconomical or object to tooling clay and they would rather have the piece completed entirely during the plastic, throwing process. The piece is then thrown so that the base is the same thickness as the walls. When throwing is completed, a wooden modelling tool is used to clean away excess clay and sludge from the bottom outside corner. The pot can be removed from the wheel head at once, to make room for the next work. A taut, twisted wire or nylon cord is pulled under the base, completely severing the pot from the wheel head. The wheel revolves slowly, and the cutting wire leaves an attractive, braided pattern under the base. The potter then deftly and gently lifts the very soft pot from the wheel with both hands and sets it aside to dry. In this method, the next step is bisk firing.

Many studios have closed cupboards or closets for storage of wheel work, in which wet pots can be dried slowly. These spaces are called *damp cabinets* or *damp closets*. If not available, a good alternative is to cover the wet piece with a large can or bucket. The clay is still so soft that it cannot safely be wrapped in plastic until it dries firm. However, many potters prefer to leave freshly thrown pieces out in the open and then, later in the day when the pots are leather hard, return to trim the feet. Left out in the open, the pot turns leather hard in a few hours. Clay under cover or in damp cabinets may take a day or two before it is that firm. That the pot should be exactly leather hard, neither softer nor harder, is a desirable condition for good trimming.

Cut the pot from its bat using a taut, braided cut-off wire or a potter's knife. Turn the wheel slowly while pulling the wire through or very gradually slide the knife blade under the pot's base. Place the piece on the wheel head rim down, that is, upside down, as close to center as the eye can tell.

There are a couple of good methods for precise centering. Turn the wheel slowly, and note on which side of the rotation there is more wobble. At that point, tap the pot sharply with a finger, nudging it closer to true center. Several more nudges should bring the top into exact position.

A second centering method may be easier for beginners. Having placed the pot upside down on the wheel head and as close to center as the eye can tell, rotate the wheel slowly. Hold the point of a tool—a modelling tool, knife, needle, or pencil—very steadily against the upper wall just under the pot's base. Where the point bumps against the revolving pot it leaves a mark. Nudge the pot away from the mark towards center. Continue to make these small adjustments until the piece is perfectly centered.

Three small clay lumps are used to hold the centered pot from moving on the wheel head. They should be firmly placed and equidistant around the rim.

Hold a flat wire, loop end tool in the right hand, as if holding a pencil or pen, but close to the working loop. Hold it at an angle of about 45°. Begin trimming across the base from the center to the periphery, then reverse. The speed of the wheel should be moderate, not slow. Cut away excess clay to make the base flat and level. Ideally, when the clay is properly leather hard, the trimmings roll away in a constant stream, in a curlicue.

Next, trim the footing to shape. The footing is in the pot's silhouette, that part of the bottom wall just above the base. Superfluous clay is cut away to make the piece lighter, and, even more importantly, to determine the foot's shape. At first, it may be awkward to visualize the pot's shape while it is upside down, but it is nevertheless vital that the footing should be realized as an integral part

7-24 *A braided wire is used to remove a leather-hard pot from the bat.*

7-25 *Cutting the leather-hard pot from the bat.*

7-26 *Locating the leather-hard pot in true center.*

7-27 *The centered pot is held in place with three clay keys.*

7-28 *Tools for trimming.*

7-29 *Trim away excess clay to make the base flat and level.*

7-30 *Trim the footing to shape.*

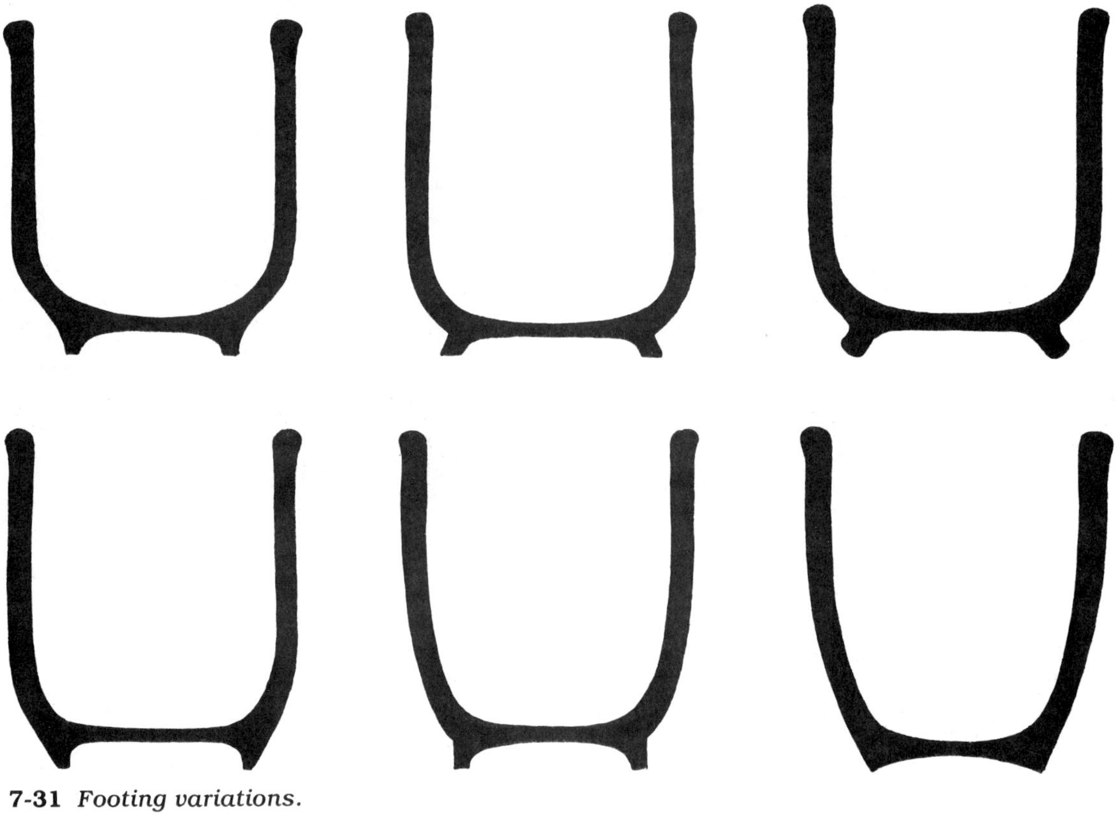

7-31 *Footing variations.*

of the form, not a self-conscious or ostentatious attachment. The footing achieves real meaning as an expression of the linear rhythm continuing from the flow of rim into silhouette and inner volume. After the footing has been trimmed, the next step is to hollow out the inside of its base.

Trim about half of the base thickness, which had been left at about 1/2". Leave enough at the periphery for a footring. Beginners may be a little nervous about going all the way through the base, fear which rarely materializes. There is a perfectly good way to tell when the trimming has gone deep enough; but it works only when the pot is leather hard. When it seems that the base has been trimmed deep enough, stop the wheel and press down on the center of the base with your thumb. If the base thickness is right, the thumb pressure causes it to flex a bit. The firm clay submits to digital pressure as if it were a drum skin.

Another concern is the width of the footring. Actually, a very narrow footring easily supports a very large pot, and, obviously, a very wide footring has no difficulty in supporting a small piece. The width of the footring is all a matter of visual scale.

7-32 ***Hollowing out the inside of the footring.***

7-33 ***Testing for proper base thickness.***

It should appear compatible in its relationship with the mass of the pot. Even the pot's underside is a component of its composition.

SHALLOW POTS, BOWLS, AND PLATTERS

Shallow forms, such as bowls and platters, receive their general proportions during the early throwing steps. Center a low, wide mound rather than a tall one. The general configuration of the centered clay determines how wide a bowl can be thrown from that piece. Open the ball into a wide cylinder with thick walls. Now, each time the clay is drawn up, pull it wider in diameter as well. There will be increased pressure from the inside hand but firm support from the outside hand. The inside hand should always start at the center of the base, the finger tip firmly and steadily moving flat across the bottom till it meets the wall and then up, opposite the outside finger. The inside finger tips leave a beautiful, unfolding spiral finger pattern in an uninterrupted sweep. There is little else in the form of a bowl but the evident and dominant volume. An expressive volume is what a bowl is all about.

Your throwing should define the inner silhouette and give continuity to the spiral finger mark. Avoid cornering at the point where the base turns softly into the wall. A steady, strong flow of curve connects the base to the wall, almost indistinguishably. The danger is that too much clay will be left in the bottom of the wall and in the base thickness. Although all extra clay will subsequently be trimmed away, it is a waste of time and material if the throwing is too thick. One exciting challenge

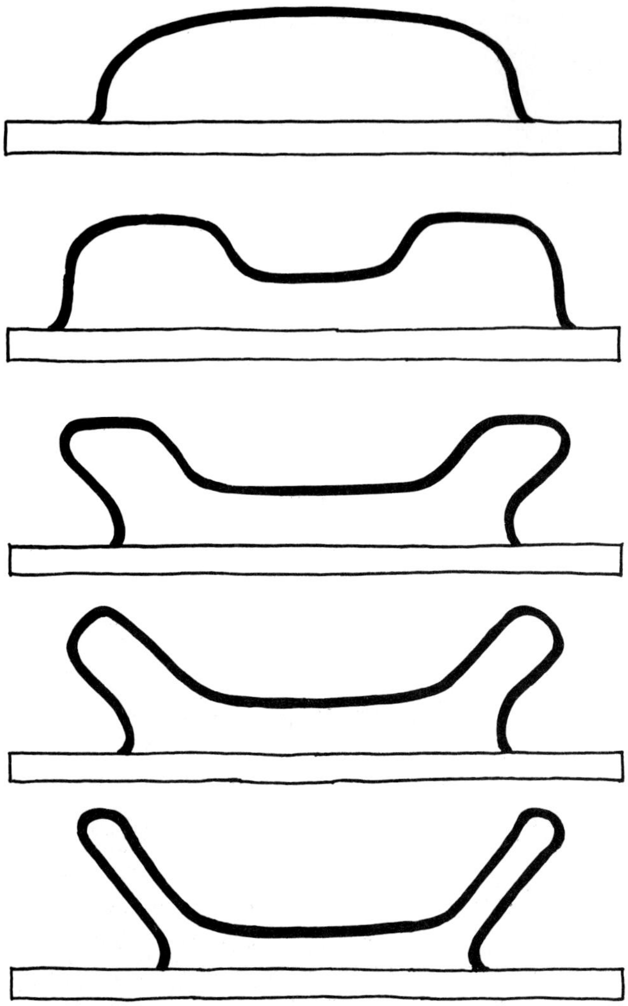

7-34 *The steps in throwing a bowl.*

in making these wide, shallow forms is to throw increasingly spacious pieces from the same quantity of clay. With practice, little clay is wasted.

BOTTLES AND JARS

Soft clay spinning on the wheel almost instinctively desires to spread out. Probably the effect of the centrifugal force, the clay wall too easily falls into a bowl shape. Most potters can recall earlier throwing experiences when all too often their pots transformed into collapsed sombreros—a consequence of the wheel going too fast and the clay being too wet. To bring the pot's wall into a vertical surge, and then to gather it into a closed neck, as in a bottle or jar, is a rewarding achievement of high order. Much practice in throwing tall cylinders should precede the attempt to make a closed neck.

A vessel becomes more rotund if the inside finger tips push out; conversely pressure inward from the outside right hand tends to make the diameter narrower. In each case, the clay is made by finger pressure to move from one location to another. After all, clay is not quite solid, it flows. Whether drawing the form in or causing it to expand, an ample supply of clay should be provided. If the pot is to be made narrower or wider, the wall at that level should be left thicker. But throwing is an act almost always performed by opposing pressures. Consequently, when more force is applied from one direction, the opposite fingers give support to the clay wall. Working with the force of one hand only often results in a rippled, torn, or collapsed wall.

Bottles and jars—pitchers also fall within this category of form and technique—are most often taller than wide. They function more efficiently in vertical scale and should not be squat and dumpy in appearance. To start one of these pots, throw a good cylinder a bit taller than needed for the dimension.

One good way to narrow the neck is called *collaring* or *necking.* With your two hands, encircle the pot below the neck, thumbs spread apart from fingers. The wheel turns a bit faster, and the fingers are clean of throwing slurry but thoroughly wet. Slowly tighten the grip around the neck. Simultaneously, move the hands gradually up towards the top, closing the fingers slightly as they move up. There should be steady lubrication, for the slightest pull of friction on the clay might ruin it all. Collar several times, until the jar has been sufficiently closed.

Another good method to make the pot's neck smaller uses the finger tips rather than the hands. With the fingers of the left hand stretched out and bent in a grabbing motion, come down over the mouth of the cylinder. The wheel rotates somewhat more slowly than in collaring. Let the outstretched finger tips touch the turning rim, with constant, ample water for lubrication. Place the left index finger inside the pot's mouth, and move it up against the opposing thumb, as if pinching. The wall is in between. The other three fingers gather in from the rear. Give support to the left hand with some fingers of the right hand. As the fingers of the left hand and thumb draw together, the neck of the pot will be seen to close. Repeat this clawing movement until the neck has been sufficiently narrowed. This method works well alone, but combined with collaring, the effort may be more effective.

Many bottles and jugs are quite full formed, plump, or even fat below their narrow necks. Some are straight sided cylindrical forms whose volumes gracefully and modestly recede into a diminished top. Most of the throwing to shape the body should be completed before the neck is narrowed. A throwing stick is a clever device to refine the pot's contours after the opening is too small for the potter's

7-35 *Collaring a narrow neck.*

7-36 *Finger tips narrow the neck.*

7-37 *Leather-hard bottles.*

hand. This tool is actually a wooden rib at the end of a long handle, inserted through the small hole at the top, to press out and make the final adjustments in the shape of the pot.

PITCHERS

Pitchers come in infinite variety and yet are quite similar in their fundamental features. The purpose of a pitcher—to store and pour liquid—imparts certain constraints on its design, but within these limits are opportunities for great diversity. To enhance its storage function, a pitcher is often rotund, pleasingly plump to fat. However, because of the type of liquid stored in them, some pitchers might be slender or even concave in profile. To keep the contents from spilling or icy liquids from getting warm, for example, the form is gathered in at the top. But the opening at the top must be large enough for a hand, dish cloth, or ice cubes. Pitchers generally have handles. For proper pouring, the major concentration of weight should be below the handle. The ideal pitcher form is taller than it is wide and has the approximate configuration of a tear drop.

For efficient pouring, a lip or spout must be properly shaped so that the contents will flow at a controlled velocity. It is easier to shape the pouring lip if the pot's neck, somewhat long and narrowed, flares a bit at the rim, as in a teacup. The rim will be less apt to tear during its shaping if it is made a bit thinner than the wall underneath. Immediately after throwing, while the clay is still fresh and plastic, gently touch under the rim with the left finger tips, the thumb set several inches apart from the index finger. The full length of the right index finger slides back and forth inside the neck, in the space between the two fingers pressing from the outside, gently but firmly urging the wall into a spout shape. An actual, strongly defined channel should be made so that there is no splashing over the sides during pouring. Some potters prefer to carry this channel all the way down the long neck into the pitcher's belly, emphasizing it as a canal, by pressing in on both sides with sliding fingers. This channel stretching from lip to belly may enhance the pot's appearance as well as improving its function.

It is not difficult to produce a dribble-free pouring lip. While pulling the lip into shape, draw it out and down further to form a slight pout or overhang. Gently slide a wet finger over the pouring lip, pulling it into an overhang. While doing this, support the wall on both sides with two outstretched fingers of the other hand.

Pitchers are a unique challenge to the potter, to fulfill the technical and functional requirements

7-38 *A pitcher.*

7-39 *Shaping a pouring lip.*

7-40 *Shaping a small pouring lip.*

7-41 *Pouring lip with overhang to prevent dribbling.*

yet develop vitality in the work. Among the elements available are proportion of mass to height, surge and energy of silhouette, rhythm of finger marks, crisp and certain throwing, clean details, and a handle in appropriate scale and spirit.

HANDLES

Pulling is the favorite method for potters to form handles. In the pulled handle are those fluid and plastic qualities which are so compatible to expression in handbuilding and throwing.

Pulling works best when the clay is excellently wedged and the correct consistency for throwing. The size of the clay wad is important, for much of it is used up in the handle; learn to use a big enough piece of clay without waste. Squeeze and roll the clay wad into a pear shape, and then hold the pear by the large end, small end pointed down. To avoid cracking of the clay the handle should be pulled down, rather than obliquely. A pan of water is kept beneath the work, as a ready source of water for lubrication and to catch the drips.

With one hand holding the pear by the large end, small end pointdown, the other hand, thoroughly wet, draws and milks the clay down in a series of steady strokes. Each stroke starts at the top and continues down to beyond the tip. From the start to the end of each pull, draw the fingers tighter, closer together. A taper results, thicker and wider at the top and thinner and narrower at the bottom. The major contact with the clay is made by that small area of palm between thumb and index finger, kept constantly wet, and the fingers should meet no resistance. Each few strokes turn the handle around for a more symmetrical cross-section.

Should a round cross-section be desired, adjust the hole in the pulling hand to make it round; if a more strap-like section is wanted then the fingers should compress into a more oval, flattened hole.

Many milking pulls are required to make a good handle, each one of a slow steady complete pull from the very top to beyond the end. With practice,

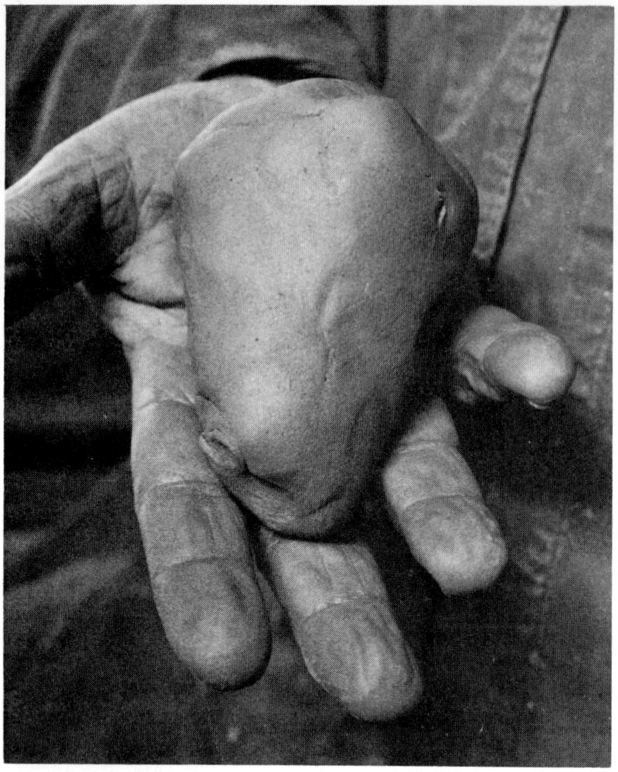

7-42 *Handles are pulled from pear-shaped wads of clay.*

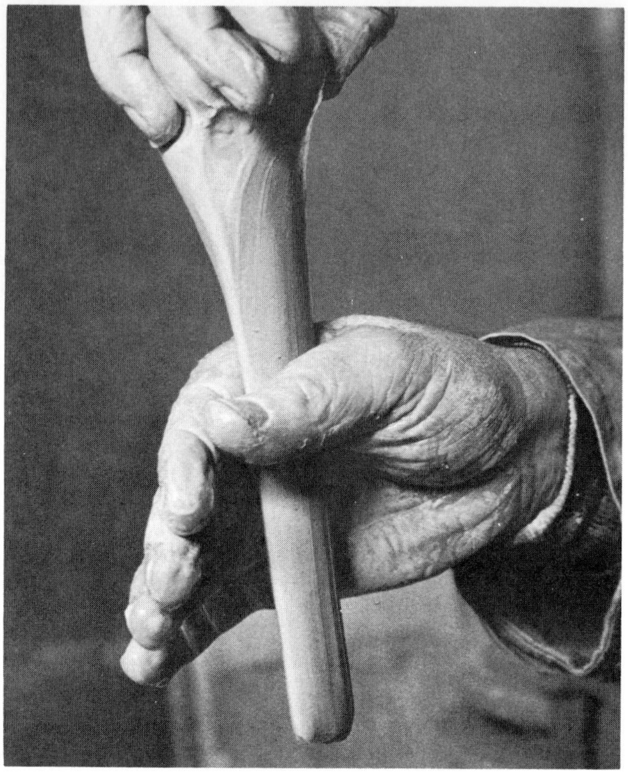

7-43 *Pulling a handle.*

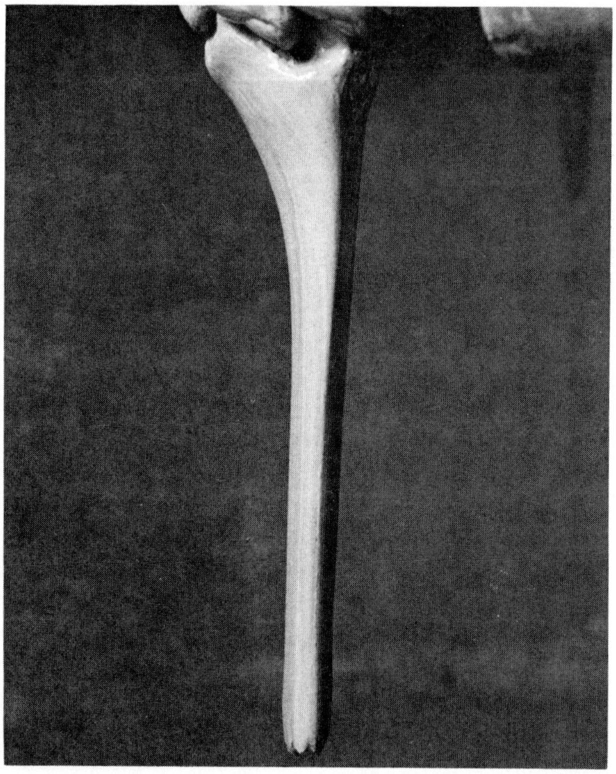

7-44 *The taper in a pulled handle.*

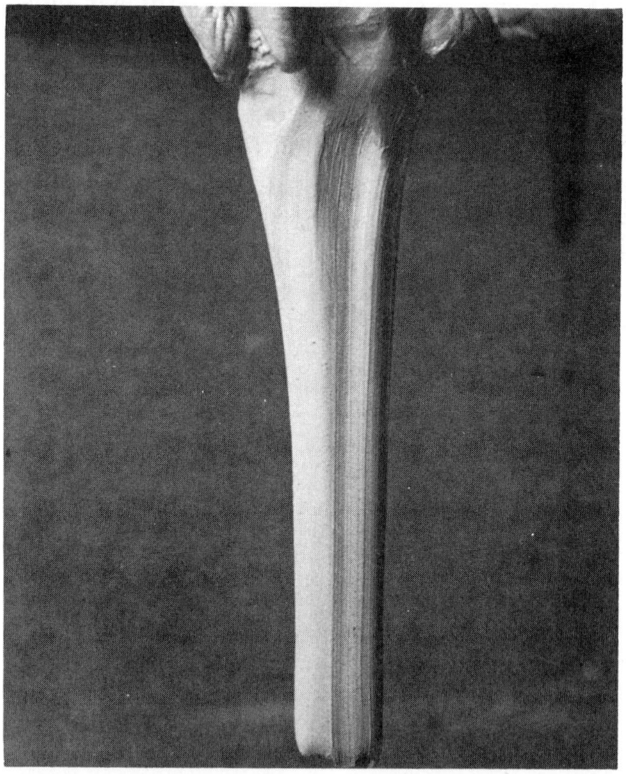

7-45 *Thicker and wider at the top.*

you will be able to complete a handle in a minute or less. It is good for the pulling to be completed in a short time, for the clay would otherwise become water-logged. Many potters have the good habit of pulling two handles for each one needed.

A suitable handle is thicker and wider at the top, thinner and narrower at the bottom—a taper. Fortunately, this fluid shape occurs naturally from good, pulling strokes. Taper reinforces the visual impact of the piece and improves the fit of the handle in the grip of fingers.

When the handle has been pulled just right in

7-46 *Forming the handle loop.*

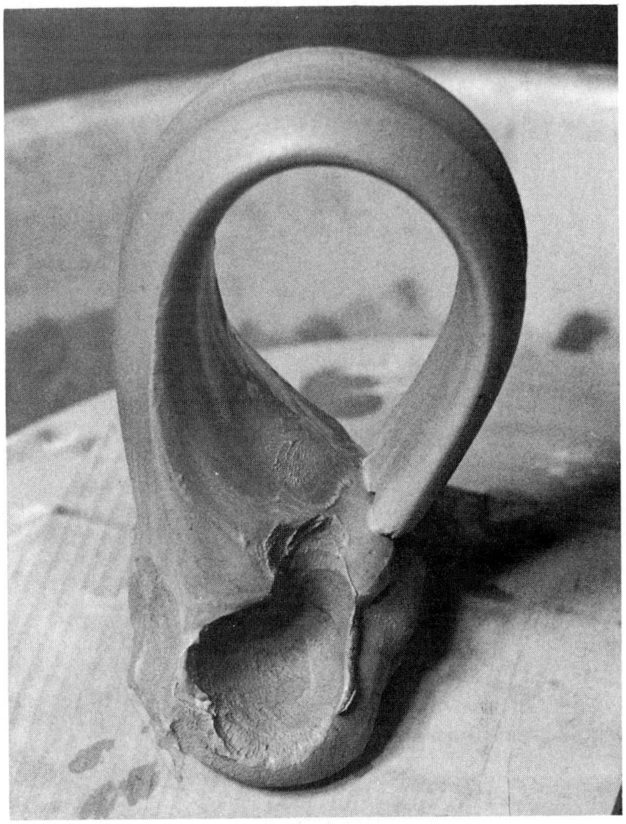

7-47 *Press the tip into the large end.*

girth and length, catch the tip and curve it back into the large end of the original pear. Press the tip lightly into place, to stick, in this manner forming the handle loop. If a handle has been pulled too long and thin, it will collapse of its own weight when curved into the loop. If the handle is pulled too short and fat, it cracks when bent into its loop. A properly proportioned taper makes a beautiful curve. Set the good, looped handle out of the way.

The water used in pulling a handle leaves the clay wet and soft, unsuitable in that condition to be attached to its pot. The handle is permitted to dry until only the surface moisture has evaporated; it may then be touched without leaving finger marks or otherwise destroying its freshness. That immaculate, fresh pulled quality wants to be retained. The surface dries in perhaps 10 to 20 minutes, but the loop remains flexible, springy, and plastic. During this period when the handle is drying, many potters use the time to trim the footring of the pot, immediately following which the handle may be attached. The pot body itself should be no harder than leather hard.

Three other techniques for handle making are (1) rolling a coil, cutting to the length of handle needed, and then, by score and slip, attaching it at both ends to the pot (The pot and handle should both be leather hard or softer for them to stick together); (2) cutting a strip of clay to the desired width and length, which is then applied to the pot by score and slip; and (3) attaching a wire loop to a wooden handle—the curve of loop furnishes the handle cross-section. Pull the tool through a clay mass, and release a continuous stream of handle. Repeat as many times as handles are needed. This method is particularly helpful when producing many handles of the same shape and size.

It is a simple matter to fasten a handle to its pot. But whether the handle will look and work well is the result of careful, sensitive thought and execution. Questions to consider include: Is the handle's character appropriate to that of the pot, in, for example, its girth, the shape of its cross-section, and the extent of its taper? Where on the pot should the handle ends be placed? How many fingers will be needed to lift the pot, and where in the loop will they grip? Will the loop be full or lean? How close to the side of the pot should the loop lie? Should the two ends be smoothed into the pot's wall, for the appearance of organic growth, or shall there be decorative tags, scrolls, and fillips at the handle ends, which give it a separate identity? Considerations of craftsmanship, function, and visual effect give evidence, in a good handle, that they have been sensitively and skillfully integrated. The handle appears in the pot's composition as a comfortable complement rather than an intrusion, and the fingers will know unerringly how to grasp the pot.

When the handle's surface is no longer sticky wet, but the clay remains springy and plastic, cut it to the desired length with a potter's knife. Sever the handle at the large end of the taper, leaving the loop length and tail intact. It is essential that the freshly pulled surface remain undisturbed. To avoid messing the handle surface keep the fingers dry and dusty, and touch as little as possible. Locate positions on the pot where the handle ends

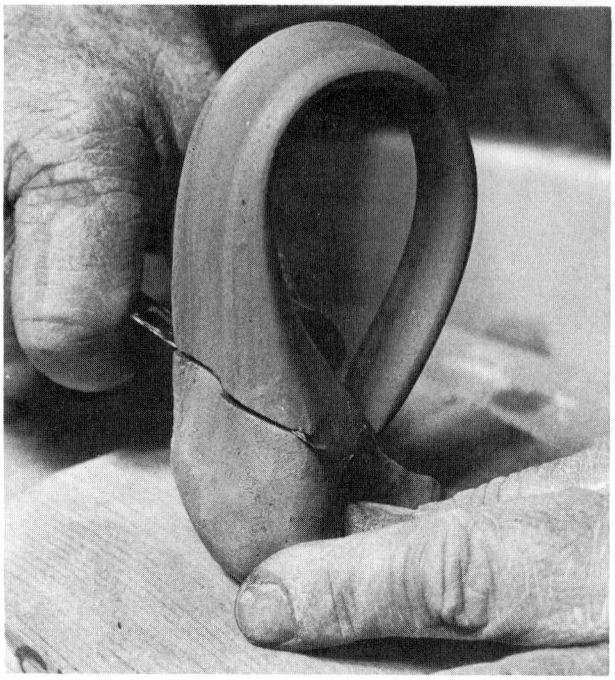

7-48 *Sever the handle at the large end.*

7-49 *Apply slip to the pot's surface.*

7-48–7-52
JOINING THE FINISHED HANDLE TO ITS POT.

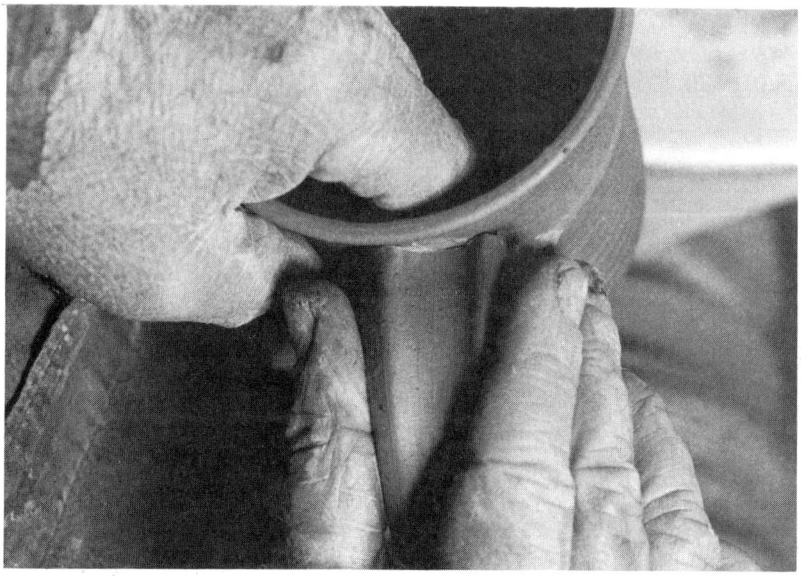

7-50 *Press the handles firmly in place. First the top . . .*

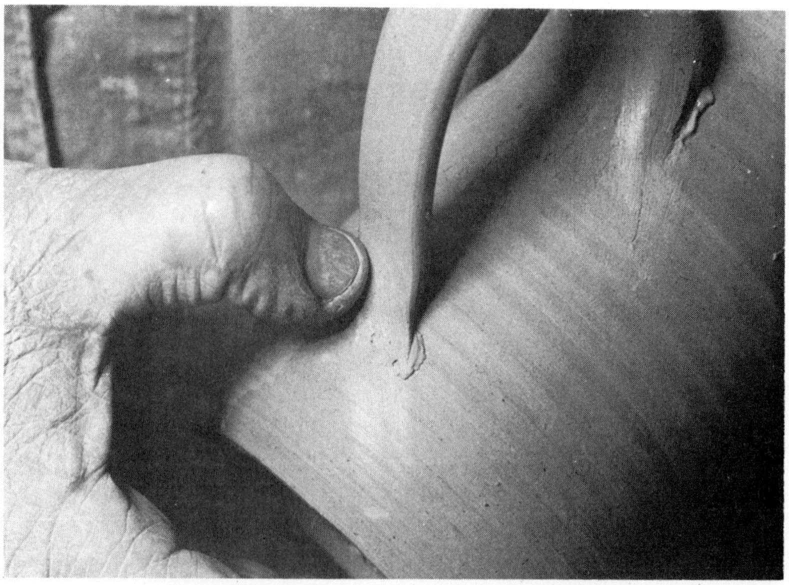

7-51 *. . . then the bottom.*

7-52 *Modify the size and shape of the loop, if desired.*

7-53 *Variations in the tag ends of the handles.*

are to be fastened. Apply generous portions of slip, and press the ends firmly into position. The top is done first, then the tail end at the bottom. At this time, adjustment and modification of the loop size and curve may take place.

The clays and the joints solidify more securely if, before being set out to dry, the pot with its newly attached handle is kept under cover overnight in a damp cabinet or covered with a plastic bag. Some potters drape a small piece of slightly moist cloth or a strip of plastic sheet over the handle portion. The cover prevents the handle from drying and shrinking more rapidly than the pot, minimizing risk of cracking at the joints. After this preliminary drying, the joints may be cleaned and freshened, as required. Now the pot may be thoroughly dried.

LIDS

A properly fitting lid sits snugly on its pot. Most lids have handles. A good lid may be lifted and put back in place without great effort. But we can so easily become concerned about the physical fit of the lid that its visual references escape adequate attention. A lid that looks right functions properly and also captures in its own identity the sense of the pot's form. It does not have the appearance of an awkward addition. It may be generous or modest in its size and dynamics, but it should promote consistency and integrity in the visual composition of the whole piece.

The cover is a part of its pot, yet often the uncovered pot is successful in its own right. Consider that a pot with lid may, in addition, have a handle or two and a spout. Each of the parts has been made separately, at different times, but they come together to accommodate each other's existence and to acknowledge functional priorities.

The primary reason for having a lid is its usefulness: It covers. It may very well be decorative and an exciting factor in the pot's composition, but above all it should be made well and work well. The major requirement in the physical fit of a lid is that it should securely cover the mouth of the pot. Too tight a fit, and even a small amount of warping during drying or firing will destroy its usefulness. On the other hand, a lid that fits too loosely is not only an embarrassment to the potter's skill but also invites the clay to chip and break. For a good fit, the lid is made from the same batch of clay as the pot, giving assurance to consistent shrinkage. Many potters prefer to throw the lid immediately after throwing the pot. There is an enhanced opportunity for integration of feeling. A set of calipers is a very useful tool for properly measuring openings and diameters, but a ruler or a straight stick marked with measurements also may be used.

Potters have a variety of techniques to shape lids for good physical and visual fit. There are three types most often used: (1) One, simple in cross-section, sits on a shelf or *flange* provided for it within the pot's walls. The shelf, thrown when the pot was thrown, prevents the lid from falling in or sliding off. (2) This type has a shelf arranged in the lid's shape. The shelf is thrown when the lid is thrown. (3) The third type is shaped like a plug. It fits down inside the mouth of its pot but cannot fall in because its major diameter is larger than that of the pot's opening.

None of these types is inherently superior to the others. The choice is based on certain considerations: whether the vessel will hold liquids, hot or cold, powders, or solids; whether a shelf within the pot might be an unsanitary trap for food residue and dirt; whether there will be some provision for a ladle, spoon, dipper; will the lid be wide, narrow, flat, domed; is the lid lifted often or is the pot used mostly for storage; will the pot be tilted in use. Not the least of these factors is the appropriateness of the lid to the pot's form.

There are no special lid-throwing tricks or exotic ways to hold fingers and use tools. The sequence of steps is the some as for any other kind of throwing, except for making certain that the clay for the lid is the same clay used to make the pot. The centered shape you start with is squat and thick, the general configuration of the lid.

Lids Resting on "Shelves"

To throw the pot with a shelf arrangement built into the inner wall, raise the cylinder, leaving a little extra thickness in the wall near the top, at the location of the shelf. Rotate the wheel more slowly. Slowly pinch the extra wall thickness inside the wall between the left thumb and forefinger, first to make a ridge, then to define it more clearly as a shelf. A modelling tool may be used to make the shelf surface and its corner with the wall more precise. With the calipers record the inside diameter, corner to corner, at shelf height. A conscious decision should be made as to how deep the shelf level is to be below the rim. Different positions will evoke different visual responses.

Either a flat pancake lid or inverted bowl shape may be used for the shelf arrangement described above. The outside diameter of the lid is the same as the pot's inside diameter measurement taken with calipers. Throw the flat lid with variations or corrugations in its cross-section. A series of shadowy, concentric rings results, to visually relieve the monotony of flatness. The variation in cross-section also helps to prevent warping during drying and firing.

The lid's handle may be thrown together with the lid, all of the same piece. Leave a small lump of clay in the center, and at the end of throwing the lid, shape the lump into a handle, the shape of a diminutive pot or knob. The handle's form is inspired by the pot's form, to complement and perhaps add a touch of excitement. The lid remains on its bat until bone dry. No trimming is needed. When completely dry, the lid lifts easily off the bat, and the underside is quickly scraped clean with potter's knife and sponged. The bowl-shaped lid is thrown in the sam_ manner as any bowl. When leather hard, the bottom must be trimmed to shape and a handle furnished.

Lids with Flanges

A lid with its own flange is thrown upside down. The throwing technique is very much like that for a bowl, except for additional thickness left in the

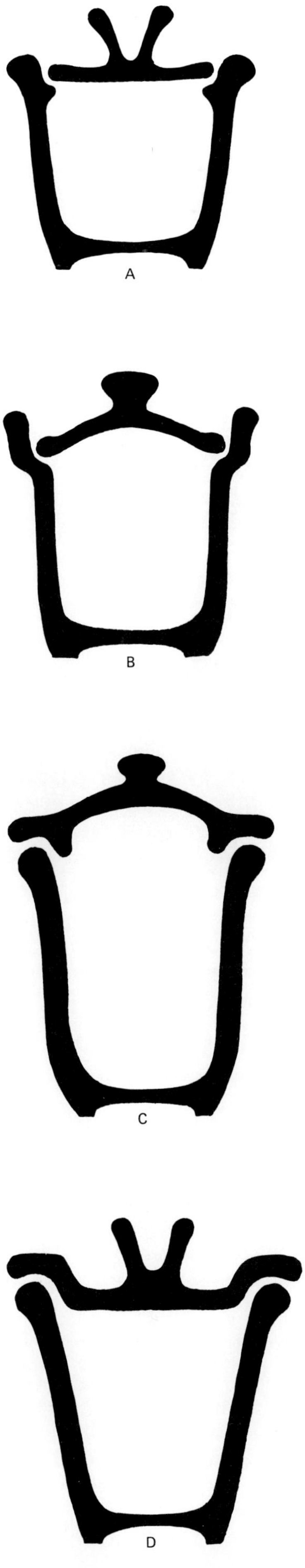

7-54 *Lids. A. Flat lid on shelf in pot's wall. B. Domed lid on shelf. C. Domed lid with its own flange. D. Plug lid.*

7-55 *A shelf in the pot supports the lid.*

7-56 *A plug lid (left) and a lid with its own flange (right).*

7-57 *Trimming the top of a domed lid.*

7-58–7-59 *THROWING A KNOB HANDLE.*

7-58 *Attach a ball of clay and throw a handle shape.*

7-59 *The finished handle.*

wall at the lid's rim, where the flange and collar will be shaped. It is not a difficult throwing technique to bend the wall around, to form a recess which fits inside the pot's mouth, and on which the lid sits. A modelling tool may be used to define the flange and its corner more precisely. The caliper's measurement of the pot's inside diameter is the dimension for the outside diameter of the lid's flange, corner to corner. Except in teapots, the lid's collar need not be very long, perhaps 1/4″ to 1/2″.

(Teapots are tilted for pouring. Therefore, to prevent the lid from falling off the pot while pouring, its collar is made deeper.) The lid is permitted to dry to leather hard, after which it is trimmed and a handle attached.

Small, pulled handles may be attached to lids, perhaps wound around or intertwined pretzel fashion. Handles may also be thrown on the leather-hard piece. Immediately following trimming, while the lid remains centered on the wheel, score, slip, and weld a small ball of clay onto its center. With the finger tips and just the slightest amount of water necessary for lubrication, proceed to throw the ball into a handle shape—a solid or hollow knob. Alternately, throw the handle separately, and when it is firm cut it from the bat and weld it by score and slip to the lid's center. This alternative is second best.

Plug Lids

The plug lid is thrown all in one piece, handle and all, on a bat. When it has become bone dry, scrape and sponge the underside clean. There is no need for trimming or attaching a handle. By virtue of its plug-like configuration, it fits down inside the pot's mouth, hanging suspended from its flange resting on the pot's rim. It seems always to be at least in part a series of shadows and light contrasts. A generous rise in the height of the handle may contribute to this type of lid a sense of majesty, verve, and vitality.

TEAPOTS AND COFFEE POTS

Completing a good teapot is a considerable challenge. Many different parts, made separately, must all come together at once to become an integrated whole. It has a sharply defined function, but its efficiency is but one component in drinking tea. For many tea drinkers, the teapot is an important expressive feature; the vessel's personality contributes immeasurably to the taking of the beverage.

Throw the pot first, followed by the lid. The teapot's mouth always wants to be constricted in diameter, to assist in keeping the contents hot and to support visually the fullness of form below, be it plump, rotund, or fat. The lid should have a deep collar, so that it does not fall off while the pot is tipped for pouring. Handles on the lid and pot are arranged to protect fingers from the hot clay walls.

The pouring spout is thrown next. It is thrown in the shape of an inverted funnel but curvilinear in silhouette rather than straight sided. Shape the spout using the same throwing techniques as when drawing in the top of a narrow necked bottle, described earlier in this chapter. It is not difficult to shape the spout's rim for dribble-proof pouring. One good method is to throw the rim thin and flared slightly in the manner of a teacup. One side of the soft rim may be pulled into the merest suggestion of a pitcher's pouring lip. Some potters prefer to cut across the end of the spout with a cutting wire or sharp knife, to bevel and sharpen

7-60 *The parts of a teapot ready to be assembled.*

7-61 *Pierce a series of holes through the wall.*

7-61–7-64 *MAKING THE STRAINER AND ATTACHING THE SPOUT.*

7-62 *The holes should have an orderly arrangement and be clean of clay.*

7-63 ***Press the spout into the slipped surface.***

7-64 ***The joint must often be modelled and cleaned for smooth articulation.***

the cross-section. The sharpened edge cuts the flow of tea at the proper moment.

Attach the spout after the pot's foot has been trimmed, but while it is still leather hard. The spout should have been left to become firm, but not quite leather hard. Some plasticity remains. Where the funnel shape is attached to the body of the teapot, the wall is punctured for liquid to come through, creating a strainer. (The entire section may be cut through in one big hole for a coffee pot.)

To make the strainer, pierce a series of holes through the wall. Use a small, round, pointed stick, a drill point, a twirling potter's knife, or even a strong straw. The arrangement of holes for the strainer should be orderly, for the sake of good craftsmanship and enhanced appreciation. Clean the holes thoroughly of crumbs and wisps of clay, particularly around the edges. Glazing will be made easier, and an undisturbed flow of tea will be permitted through the holes.

The spout can now be welded into place. Score and slip the wall around the hole area. Cut the spout from the funnel shape, and press it firmly into the thick slip. The joint must often be modelled and cleaned for smooth articulation. Locate the spout's tip near the level of the pot's rim. In this manner the teapot can be filled to capacity, without the embarrassment to the potter of tea prematurely flowing from a spout which has been placed too low.

Pulled handles attached opposite the spout work well for teapots, and so do bail handles, which are constructed like the handles of pails. Bamboo seems to be the most popular material for bail handles, which are available in a wide variety of sizes, shapes, and finishes. A bamboo handle is attached to the

7-65 *A teapot with a pulled handle.*

7-66 *A teapot with a bamboo bail handle.*

pot after the final glaze firing, by having its ends hook through loops on the pot's shoulders. Make the loops of coils or pulled strips. Weld them opposite each other on the leather-hard teapot, following the spout's attachment. This is the final step.

During the assembly of parts there will have been much touching, fingering, and handling of the clay surfaces. Nevertheless, the potter is very much concerned lest the freshness and immediacy of throwing has been lost. This thrown quality is irreplacable. A respected teacher has warned, "The marks of torture should not show." The fingers should always be kept dry and clean, wiped frequently, and kept dusty so that they will not mess the textures of throwing.

8

CASTING SLIPS AND DECORATIVE ENGOBES

Slip, or liquid clay, is a very handy material to have around the pottery studio. It is the "glue" for welding together two pieces of clay—score and slip—such as coils, slabs, and handles to pots. For this purpose, some potters prefer to use specially prepared slip, or *slurry*, a by-product of throwing. Slurry is really a kind of uneven, unrefined slip. Some potters would even rather throw with slip than with water. Some slips make excellent glazes, for they melt into a glass at a practical, low temperature. Large numbers of duplicate pieces can be mass produced by *slip casting*. Slip is poured into plaster of Paris molds, in which the slip hardens to assume the shape of the mold. This method is mostly used in pottery factories, but it may also be practiced on a smaller scale in pottery studios. Lastly, slips may be applied for decorative purposes to clay surfaces. When slips are used for decoration, they are sometimes referred to as *engobes*.

ENGOBES

Probably the simplest way to put engobe on clay is to paint it with a brush. The slip can be laid on as a line, or it may cover an entire area. Engobes may be applied by daubing with a sponge or rag. Slip can be sprayed on the pot's surface in a fine mist, or it can be squeezed through the small hole of a container, such as a ketchup dispenser or ear syringe. This leaves on the surface a long, raised trail similar in some respects to cake decorating. Thin layers of slip can be built up on top of each other by brush, until a raised relief results. The pot may be dipped into a bath of slip for total coverage, as when making a jelly apple. And slip can be used to fill in carved, incised, or scratched marks, or marks stamped into the clay wall. (These and other decorative techniques are explained and illustrated in Chapter 9.)

The potter should always understand slip for what it really is—a clay or pottery body. It behaves in characteristic clay fashion in its reactions to being wet, to drying, and to the heat of the kiln. Slip shrinks during drying and firing, it becomes permanently hard as a result of firing, and like any clay or body it melts into a glass if fired high enough. But because it is used as a coating, it has certain unique traits. Above all, engobe must persist as a coating; after all, it is seen because its color differs from the pot's color. Consequently, the requirements for a good slip are that it must cover, adhere, harden during firing at a lower or the same heat as the pot; it must not be *fugitive* (dissolve in reaction to a glaze); and it may not bleed or run under a glaze.

The best time to apply slip is when the pot is leather hard or softer; for enduring adhesion the softer the pot the better. The two clays, that of the pot and the slip, will be more likely to bond and shrink together in mutual adjustment. Slip may crack and flake off when applied to bone dry or bisk pots, a consequence of disparate shrinkage rates, and of the pots themselves having already shrunk. But engobe recipes are available for application on bone-dry and bisk wares. One is given below. It will be seen that this type of slip composition contains filler materials to control drying shrinkage and flux materials to fuse the slip during firing with the clay. (See Chapter 4 for a discussion of this problem.)

It should not prove difficult to find a clay good enough to use as an engobe. Some common clays are very suitable, combined with enough water for the proper consistency and applied in one of a variety of techniques. Their iron-rich color contrasts effectively with the color of many pottery clays, such as white earthenware or grey stoneware.

However, white, light, and other than iron-colored engobes must usually be mixed from a formulated recipe—rarely in nature are there white or colored clays, other than common red clays, which alone make satisfactory engobes. They hardly ever fulfill all the requirements for a successful slip. It would make good sense for a potter to have a basic, white engobe recipe, which he could then color as he wished. Two recipes are given below. In addition to the clay content, the list of ingredients contains fluxes to harden the slip during firing, filler materials to control shrinkage, hardeners to stiffen the unfired slip, and opacifiers for the slip coating to remain visible. These ingredients are balanced well for good application and retention.

The following sample recipes work well over a broad firing range, cones 04 to 9:

WHITE ENGOBE RECIPES

Leather Hard or Softer		Bone Dry	
Kaolin	25	Feldspar	9
Ball clay	20	Ball clay	16
Flint	30	Kaolin	8
Feldspar	17	Flint	11
Whiting	2	Borax	3
Magnesium carbonate	6	Zircopax	20
		Frit 3124	18
	100	Nepheline syenite	12
		Whiting	3
			100

Colorants that may be used in the above recipes are listed in the next chapter.

The potter decorating with engobe realizes early on that the stuff is not paint or paint-like, not comparable to oil paint, acrylic, or watercolor. There is a characteristic clay-thick dimensional quality to the slip, which is most evident when a coarse or stiff bristled brush is used to stroke the engobe onto the pot's surface or when it is trailed from a syringe in an oozing, continuous, raised welt. Engobe remains a distinct layer of material on the pot, whether a brush stroke or a raised line. Being clay, it dries and fires in place.

Some very effective slip decorations are, to a large extent, aesthetically successful because of the manner in which the slip appears as an extension of the pot's form, rising out of the clay rather than appearing as a superficial, applied layer, foreign and separate. Surprisingly in these good compositions, where form and surface are in harmony, color does not dominate the visual arrangements. Rather, the interplay among the many levels of tactile and visual sensations furnish the major excitement. On some pots, we see engobe applied in the same color as the pot, but it is applied thickly to evoke tactile appeal and enrich form subtly but none the less effectively.

SLIP GLAZES

Slip glazes have been used often in the history of pottery. A slip glaze is a glaze composed largely or entirely of clay. It is called a "slip" glaze because the mixture is applied to the bisk pot in the form of a slip. Many common red clays can be made to work well as this type of glaze. The clay is first refined. Fluxes may then be added so that the clay melts at a temperature lower than the maturing temperature of the pot. Consequently, it is more practical for slip glazes to be used with high-fired pottery, where the temperature usually exceeds 2150°.

The visual qualities in these glazes result from the melting behavior of the clay's components. The characteristic color is dark, iron-rich rusty or plum browns, sometimes tending almost to blacks. In texture they are thick, sometimes unctious, and shiny. Slip glazes are usually very dependable glazes. Probably the best known slip glaze clay is Albany slip, named after its source in the upper Hudson River valley near Albany, New York. Good examples of slip glazes may be seen on the shiny, dark, plum-colored glazes on early American jugs, crocks, and telephone-pole insulators.

TERRA SIGILLATA

Terra sigillata is a super-refined slip. Used extensively in ancient Greek and Roman times, there has recently been a renewed interest in this material. Traditionally, a common red clay in the form of a well-mixed slip is permitted to settle until only the finest particles remain suspended in the water, the coarser particles having sunk to the bottom. The suspension is separated from the remainder by decanting, and this is the material called terra sigillata. The thin slip can be easily applied to pottery surfaces.

Sigillata is not a glaze in the sense of a melted,

glassy coating, but it does somewhat improve water resistance at the pot's surface. The clay particles are so fine and, therefore, so densely packed as to present a more impenetrable layer. However, the contemporary appeal of terra sigillata seems to be the tactile and visual interest of this ancient pottery material, rather than any functional advantage it may contribute. Not quite glaze, but more than engobe, it presents a dull but smooth, close surface with a broad range of iron-influenced hues, compatible with many contemporary directions in pottery form.

SLIP CASTING

In the time frame of ceramic history, producing pottery wares by pouring slip into molds is relatively recent. Probably the earliest molds were made of soft bisk, extremely porous and absorbent. Today, slip molds are invariably made of plaster of Paris. Plaster is convenient and practical for this purpose, as it is easily shaped in its liquid and cheesy states. The nature of plaster is to harden quickly within about 1/2 hour of being mixed with water. The mold is permitted to dry, after which it is suitable for slip casting. A dry plaster mold is considerably porous; it can absorb up to twice its own weight in water. This feature makes it useful for casting with slip. Molds can be used many times over to produce large quantities of exactly the same shape. The molds eventually wear out. (Plaster in the pottery studio is described in Chapter 17.)

A simple mold consists of one piece. Intricate pottery shapes are cast in molds made of several sections, the parts interlocked to hold them together. These multi-piece molds are tightly bound around with cord or elastic bands during the casting process. Some very involved pottery forms and sculpture are cast piece-meal from several molds. The castings are subsequently assembled with slip.

Hollow pieces, such as most cast utilitarian wares, figurines, and small sculptures, are drain cast. The mold is filled with slip, and after a suitable period during which the clay wall forms inside the mold, the excess slip is poured, or drained, from the mold.

A good slip is creamy or syrupy in consistency. Dip a few fingers into the slip; when the fingers are withdrawn the slip should momentarily form a web between them before slowly dripping down—a practical test for proper condition. If needed, the slip may be put through a wet, fine screen to remove lumps and particles.

Pour the slip down into the center of the dry mold, in a steady stream without splashing. Avoid pouring on the sides of the mold. Fill the mold to the top. Immediately, the porous plaster absorbs water from the slip, and the slip level sinks down into the mold. More slip should be poured in to the top of the mold, and this full level kept constant until the mold is drained.

Test the casting's thickness. Use the edge of a potter's knife to scrape a small section at the top edge of the casting. A skin of solid clay will have formed around the mold's casting surface. When more water is absorbed, the casting becomes thicker. A good wall thickness sets in a few minutes. The casting's exterior surface follows the configuration of the mold precisely, and the wall thickness of a drain casting is exactly uniform throughout. When the casting wall has attained the proper thickness, the mold is turned over for the excess slip to drain.

Drain the mold into a catch basin underneath, to capture the reusable, extra material. Turn it from side to side and tilt it some in a turning spiral for the slip to flow evenly from the mold. This should help get a good, smooth, uniform inner surface, devoid of dribbles and layers. Refrain from shaking the mold up and down to assist in draining, as this usually results in an uneven, layered interior.

8-1 *A multi-section drain mold.*

8-2 *A one-piece drain mold. From left to right: the model around which plaster was poured to make the mold, the mold, two castings.*

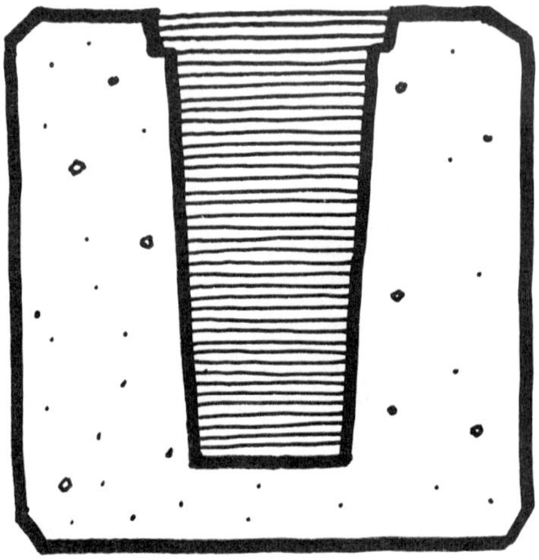

8-3 *Fill the mold with slip to the top.*

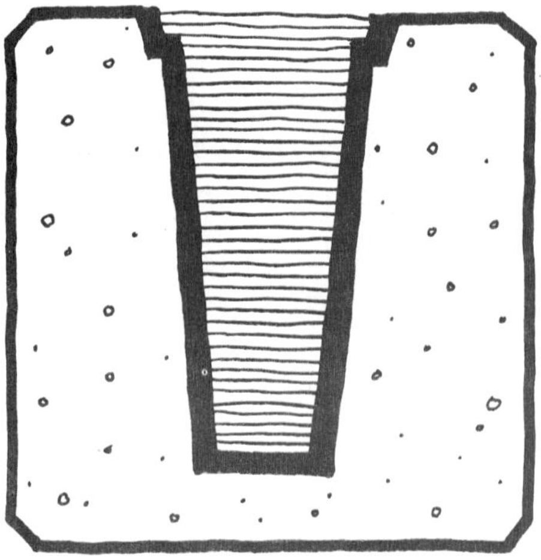

8-4 *A wall solidifies around the mold's casting surface.*

8-5 *When it is ready, the casting frees itself from the mold.*

Permit the mold to sit undisturbed, till the casting separates from the mold surface. In a few minutes it dries sufficiently for shrinkage to draw it away. The casting may then be easily removed, a task made simpler if the mold has been sitting upside down, supported by a couple of sticks. The casting falls out of the mold when it is ready, rim first.

Often, only the rim need be finished, by careful sponging with an elephant-ear sponge. If the casting has come from a multi-piece mold, it will be necessary to scrape with a potter's knife and sponge clean the parting lines that appear in the piece where the mold parts had joined.

There is another method of slip casting, in which the casting slip is permitted to solidify completely within the mold. In this way, pottery forms can be cast whose entire surfaces and cross-sections are determined by the mold. In solid-cast wares the wall cross-sections are varied in thickness, as opposed to the characteristic uniform wall sections in drain casting. Molds for solid casting consist of two or more sections; drain molds are usually unsuitable for solid casting.

When a plaster mold is dry, it may be used repeatedly until the mold becomes saturated. The evidence that the mold is too wet is when castings stick, or they come free very slowly. Castings separate more easily from a mold if the plaster surfaces are first dusted with potter's flint. With a soft brush apply the powdered flint to the inner surfaces, then blow away the excess. Between uses and after the plaster has become too wet, the mold should be permitted to dry. Set it on some sticks for maximum air circulation. Keep it away from high and direct heat, such as the top of a kiln, for the plaster deteriorates under those conditions. A good mold can be expected to have a life expectancy of some 50 to 75 castings, after which it may become so inefficient that it must be discarded.

CASTING SLIP

Some clays and pottery bodies function quite well as casting slips. A handbuilding or throwing clay might, if not too coarse textured, work satisfactorily. Consequently, prior to purchasing or formulating an original casting mixture, try the clay already on hand. A variety of commercially prepared slips for casting are available which may be purchased by weight or by liquid volume. They come in a range of firing temperatures and colors. Test a sample in your studio, before investing in a large quantity.

A good casting slip has the following qualities:

1) It will remain stable as a creamy, uniform liquid for a long period; it will not settle, separate, or thicken.
2) It pours without lumping.
3) It does not wet plaster molds prematurely—the mold can be cast several times before it must rest to dry.
4) Castings release quickly from the mold.
5) Castings shrink within permissible limits.
6) Castings do not crack or warp during drying.
7) There is adequate dry strength for the piece to be safely handled without undue breakage.

This is a long list of requirements. Many clays and bodies will need adjustment in one direction or another. Fortunately, the major adjustments are not difficult.

The same attention must be given in the formulation of a casting body to the properties of plasticity, porosity, and vitrifiability as for any other pottery body. These properties are fulfilled by balancing the relative proportions of plastic materials, stiffening agents, and fluxes. Read Chapter 4 for a description of the theory and the process. The materials properly chosen for casting will not only achieve a good mix of flow, shrinkage, and firing characteristics but will also adjust the casting body for a particular firing range.

Probably the most difficult problem in formulating and using a slip-casting body comes from the excessive wetting of the plaster mold. The major consideration is to have the use of the mold for many castings, without the necessity of it lying idle for drying too often. Related to this is the need for each casting to release quickly and easily. When a casting sticks in the mold even temporarily, not only does that waste time, but there is the likelihood of warping during drying. These and other casting difficulties point to one condition: the proportion of water in the casting slip.

Ordinarily, when a dry-casting mixture is combined with water to form a slip, the particles of clay arrange themselves at random in groups or *flocks*. As much as 50% water will be used to obtain a good, creamy, free-flowing fluid. It is a very large amount of water and is the cause of excessive wetting of the mold. Castings stick, shrink excessively, and tend to warp. The remedy is clear, and potters can reduce the amount of water needed by the slip, and in this way improve its working properties.

An electrolyte is added to the slip. This chemical is called a *deflocculant*. It causes clay particles to repel each other. The particles disperse more uniformly and require less water to go into a good slip. The deflocculants usually used are sodium silicate (water glass) and soda ash. They may be used together or each separately in a proportion of one percent or less of the weight of the casting body. Most commercially prepared casting slips are deflocculated, but a few may not be. This information will be noted on the package.

Should it become desirable to adjust a purchased body, a slip composed of casting scraps and trimmings, a studio clay, or an original formulation, the ideal relationship of water to dry ingredients and deflocculant must first be determined. A good procedure for this follows:

Measure 200 ml. of water and place in a bowl with a round bottom, about 8″ in diameter and 4″ deep. Slowly add 200 grams of the clay mixture, while the powder slakes. When the clay has been thoroughly absorbed, stir vigorously by hand until the clay is completely dispersed in the water to form a uniform slip. Depending on the clay being tested, the slip might be watery or too thick. If the slip is too thin for good casting, add 50 grams or more of the clay until any further addition would make it too viscous. Record how much clay has been added.

Now add 1 ml. of a solution of 50% sodium silicate and 50% water, and stir in thoroughly. Observe the effect of the deflocculant on the viscosity of the slip. In most cases, the slip has been thinned, and more dry clay can be added until it is transformed into a satisfactory casting consistency.

The procedure of adding increments of sodium silicate solution and dry clay is continued until the potter determines how much is needed in total to achieve the maximum effect of deflocculation. Without adding more water, try to work as much as feasible of the 500 grams of clay into slip form.

After the test has been completed, set the bowl aside for 10 minutes and note whether the slip thickens or gels. A clay that gels will cast more slowly, while a non-gelling slip casts more quickly. A slip that gels should have measured amounts of water added to it until it is adjusted so that it will no longer gel. This is actually a reversal of deflocculation.

From this test the proper proportion of water, clay, and deflocculant can be calculated, and the information used to mix a large quantity of good casting slip for your studio.

POTTERY
IN
COLOR

POTTERY
IN
COLOR

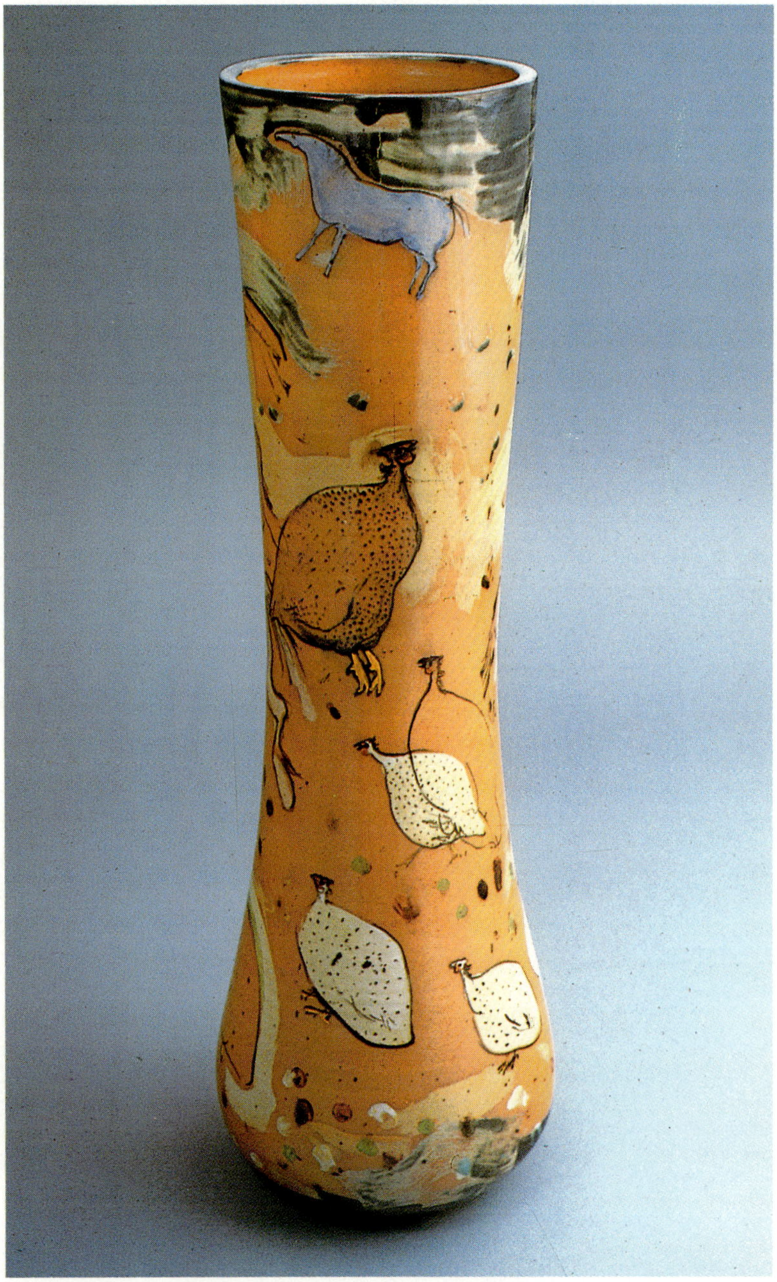

JENNY LIND. *Vase with animals, cone 10 porcelain with overglazes.*

David Shaner. *Gourd-covered jar, thrown and altered, wood-fired fly-ash glaze. 10″ × 12″.*

Karen Karnes. *Covered vessel, stoneware, wood-fired. 11½″ × 18″.*

NEIL TETKOWSKI. *"Mare de Stelle," wall sculpture, terra sigillata, fired at approximately cone 04. 38'' diameter.*

Photograph by Jamey Stillings.

SANDY SIMON. *Cappuccino cup, thrown, with surface decoration of underglaze stains and bisque-imbedded "clay rocks," fired to cone 4 oxidation, porcelain clay sprayed with a commercial clear glaze. 3'' × 4''.*

Photograph by Michele Maier.

TOM COLEMAN. *Porcelain teapot, white Chun glaze over translucent porcelain, fired to cone 10 reduction. 10'' (with spout) × $11\frac{1}{2}$'' (with spout).*

Photograph by Rick Paulson.

Frank Boyden. *Salmon vase, incised stoneware with iron, manganese glaze stains, and porcelain slips. Height: 22″.*

Photograph by Jim Piper.

Bob Green. *"Globe," pit-fired earthenware, burnished terra sigillata and glazes. 16″ × 12″.*

Wally Mason. *Slip-cast whiteware, underglaze, glaze, pottery shards, and pencil. $22\frac{1}{2}'' \times 25'' \times 2\frac{1}{2}''$ deep.*

Toshiko Takaezu. *"Moonscape," stoneware. Diameters: 28″, 23″, 20″, 15″, and 10″.*

Charles Lakofsky. *"Squash" bottle, slip-cast porcelain. Diameter: 7″.*

Ralph Bacerra. *Plate, earthenware, colored engobes, overglaze enamels, metallic lusters. Diameter: 23''.*

Graham Marks. *Earthenware, coil construction. 31'' × 31'' × 30''.*

Collection of Edna Beron. Courtesy of the Helen Drutt Gallery.

John Glick. *Boxes with lids; extruded, altered stoneware, multiple slips and glazes, fired to cone 10 reduction. The taller has a height of 4½″, width 4″ × 5″.*

Cynthia Bringle. *Raku, wheel-thrown and altered by incising and carving, white raku glaze with stains.*

N. W. Shaul. *Covered jar, stoneware. Height: 12″.*

9

THE THEORY AND PRACTICE OF DECORATION

Decoration is ornamentation, adornment, and embellishment apart from pottery form. It is usually manifested in surface, but it may be an extraordinary element of the form—a decorative knob or handle, an added fillip, or an ornate footing. Decoration constitutes all that visual material which is supplementary and accessory to the form, supporting it and contributing to its vitality. However, potters continue to debate the uses of decoration. For example, some pots are called "decorative wares" because they seem not to have any usefulness except to visually enrich their surroundings. But may this not imply that if functional or "useful" pieces have value, then "useless" pots are worthless? The reader will soon be convinced that good work of any kind enhances one's life, however the pots are used. By "decoration" we shall mean whatever about the work intensifies and gives further meaning to its form.

ELEMENTS OF DECORATION

Decorative elements are the visual fundamentals, the things we actually see and arrange on the pot's form. Arrangement is another term for "composition," the word artists prefer for its precision of meaning. Among the compositional elements available to potters are color, texture, line, and shape.

Color

Color is probably the most evident decorative element. It can be introduced into the work at all stages: The clay might be made white or colored; color might be applied as a slip or pigment in the plastic, leather-hard, bone-dry, and biscuit stages; glazes may be variously colored; colored stains may be put into unfired and fired glazes; and pigment may be applied on the surface of fired glazes.

We usually recognize the identity of a color by means of its name or hue, such as blue, black, or red. However, there are two other properties of color: how bright or dull the color is, its *intensity*, and how light or dark the color is, its *value*. The more sensitive and successful pottery decorations reflect intriguing interplay between these two qualities, rather than merely utilizing combinations of hues.

Some colors are brighter than others; red, for example, is brighter than green. Greater opportunity for more creative manipulation of color arises when the potter recognizes that each hue can be modified to make it less bright or duller than its original purity or to alter intensity. Moreover, we know that some colors are by their nature darker than others; blue is darker than yellow. But we can make a hue darker until it is black or lighter until it appears white, by modifying its value. We see a pot which possesses exquisite coloration. Upon closer examination we perceive that only one hue has been used by the potter, expressed in a myriad of rich tones solely from the range of values and intensities.

However, this somewhat theoretical description of the potter's palette is, in actual practice, considerably modified by the chemical characteristics and idiosyncracies of the materials and processes. Some hues, values, and intensities are easy to come by, while others can be generated only after difficult and complicated arrangements. Therefore, comprehension of the uses of color in clay work coincides with an understanding of its materials and techniques.

The ultimate source of all color in pottery is a variety of metallic oxides. For example, if we add copper oxide to a glaze, we shall expect it to fire green. However, the other chemicals present in the glaze and the circumstances of firing may have dramatic influences on color results. Copper oxide may generally be expected to give a green, but in some firing situations a red appears, and in other glazes turquoise shows. Other metallic oxides behave similarly in their responses either to the glaze environment or the conditions of firing. The unpredictable occurs often enough, but potters attempt to harness chance. They learn to control and use these phenomena for the good of the pot. They learn, for example, to have copper turn out green or red, whichever is desired. The precise hue, value, and intensity that ultimately appears in the pottery results from complex interactions among the clays, glazes, colorants, and firing conditions.

In the following table, oxides used in the approximate percentages indicated may be expected to give the hues listed when fired in an electric kiln or in the oxidizing atmosphere of a combustion kiln. (Read Chapter 12 for an explanation of various firing conditions and kilns.)

TABLE OF METALLIC OXIDES

Oxide	Percentage	Color
cobalt	$\frac{1}{2}$ to 2	blues
copper	1 to 4	greens
chromium	1 to 3	greens
iron	2 to 10	rusty reds
manganese	1 to 4	lavenders
nickel	1 to 4	greys
antimony	2 to 5	yellows
iron chromate	2 to 5	greys
ilmenite	1 to 2	dark speckles
rutile	5 to 15	buff, tan
vanadium	2 to 5	yellows

These metallic oxides produce colors varying in hue, intensity, and value depending on the strengths used and the composition of the materials with which they have been incorporated. The percentages shown are suggested ranges. Lesser or greater amounts may be tried. A higher percentage results in a deeper color, and a lower one gives a lighter color. For example, 1% red iron oxide in most glazes gives a light, pale, straw color, while 10% or more of the same oxide produces a deep rust to a dark brown, sometimes favorably endowed with furry streaks or oily spots. Copper and manganese in significant proportions give the glaze a black, metallic lustre, sometimes referred to as "gun metal black."

Any of the above colorants used alone gives approximately the hue indicated for it. However, two or more oxides may be combined to produce a different color. When mixing together two or more colorants it is wise to use less of each; otherwise they tend to overcome each other. A strong hue such as the raw blue of cobalt may be toned down by adding a small percentage of iron or rutile to it. Iron chromate may be made a warmer grey by adding some iron oxide to it, the contribution of its warm rust tone. There is no good source of black in a single colorant oxide; it results best from a mixture in significant amounts of cobalt, manganese, iron, and copper. The reader is encouraged to make numerous color tests—a time-consuming task, but one which is bound to result in an interesting and fertile color palette.

It cannot be stated too strongly how sensitive these colorants are to the chemical environments of their clay or glaze mixtures. Sometimes the unexpected occurs. For example, chrome oxide can be expected to produce a strong, green hue (some potters prefer the gentler green of copper). But, in a glaze where tin or zinc oxide is also present, the chromium turns tan, brown, or a variety of low-keyed red. That is one method pottery-color manufacturers use to produce red colors: They combine chromium, tin, and zinc. Obviously, it is prudent to try a small sample of clay, slip, or glaze with the oxide colorant contemplated before investing large quantities of time and material.

The following is a practical method to test the colorants that have been added to either clay body, slip, or glaze:

1. Weigh dry a large batch of the mixture. Subdivide it into an equal number of smaller batches. For example, weigh out 500 grams of a glaze recipe. When the total amount has been placed on a clean newspaper sheet, sift the dry powder through a clean, dry sieve onto a second clean newspaper sheet. Repeat the sifting four or five times, by which time the mixture is thoroughly and uniformly dispersed. Divide by weighing into five separate samples of 100 grams each.

2. Each 100-gram batch should be kept apart on a piece of paper or in a container, such as a sandwich bag.

3. For the colorant additions, weigh:

a. five different percentages of a single colorant, or

b. a percentage of five different colorants, or

c. five colorant combinations.

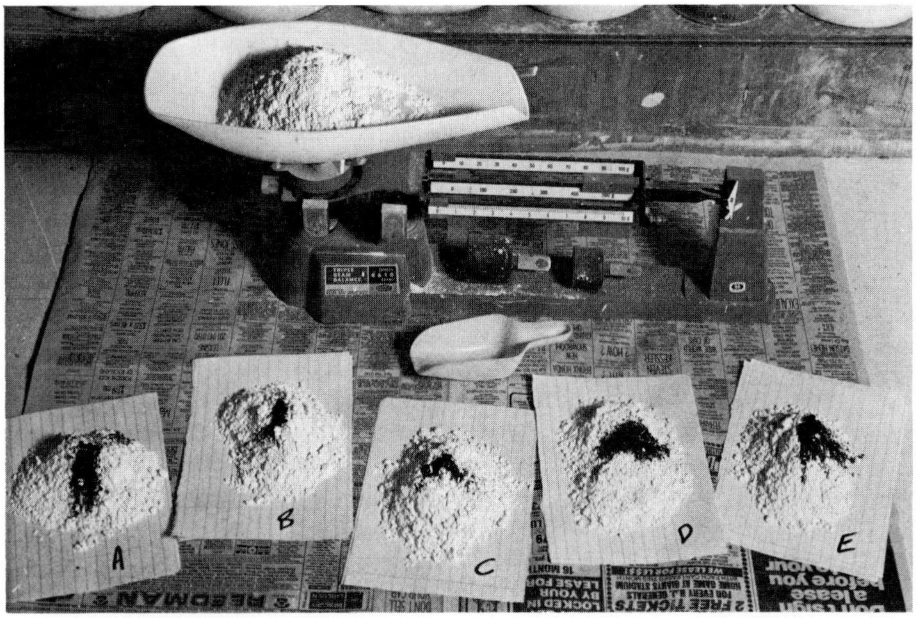

9-1 ***A five-hundred gram batch has been divided into five equal portions and a colorant added to each.***

4. Add each of the color samples to each of the glaze samples.

5. Combine with water in a mortar, and grind thoroughly with a pestle until the mixture is creamy, smooth, and uniform.

For a clay- or body-color test, pour the ground mixture on a plaster slab or bat. Permit it to dry until plastic. Shape the plastic clay into a cookie or tile for firing. Dry, then bisk fire. Glaze and fire again for a true test.

To test the color in an engobe or slip, brush the color samples on leather-hard clay. Each sample may be brushed on a separate test piece or all together, next to each other on a single test pot. Dry the test pieces, then bisk fire them. Glaze and fire again for a true test.

For color tests in glaze, apply the samples by brush, dipping, or pouring on different parts of one large test pot, or each separately on a test tile. Glaze fire.

To combine the ingredients most completely, colorant oxides are best incorporated at the time all the dry ingredients for the clay, slip, or glaze are being weighed. Oxides may be made liquid with either water or a fine oil, for application to the surfaces of bisk or glazed work. Mixed with water, the colorants may be put on unfired glaze surfaces. Color applied on fired glazes does not appear the same as color applied on unfired glazes; it may seem sharper and more distinct, whereas on the unfired glaze the colorant may bleed some and look softer.

There is a better chance to predict accurately how a colorant will behave when it is weighed and added to batches of material. However, when the oxides are casually mixed with water and brushed or sprayed on a pot, one cannot tell the outcome exactly. Color in slip and clay discloses its true character only after the samples have been glazed and fired. It makes good sense to use a transparent glaze over these tests for maximum visibility.

Mass production in the pottery industry demands both that the effect of colorants be precise and predictable and that there be a wide range of standard hues from which to choose. Therefore, a number of refined pigment compounds have been developed. They are blends of selected metallic oxides and other ceramic materials. Each comes in an almost infinite number of hues, values, and intensities. They are stable and uniform. These ceramic pigments are

1. Body stains.
2. Glaze stains.
3. Underglaze colors.
4. Overglaze colors.

Body stains are ceramic pigments added to pottery bodies in the early, dry-mixing stage. They should be evenly and completely dispersed throughout the mixture. The amount used determines color intensity, and usual proportions are up to about ten percent. These stains are also valuable to color engobes, which are, after all, clay bodies.

Glaze stains are pigments prepared specially for easy and effective incorporation into glazes. Glazes melt, while clay bodies do not. For all proportions to be accurate, glaze stains should be added to the glaze at the time the glaze ingredients are being weighed. They may also be applied as a paint to unfired glaze surfaces.

Underglaze colors are prepared blends of colorant oxides and other ceramics minerals. They are formulated for application on bisk pottery, prior to glazing. Subsequently, when the glaze melts over the underglaze decoration, the color remains in place without bleeding or running. Underglaze colors are calculated to resist color change under the attack of high temperatures. Generally, they are not as bright in intensity as overglaze pigments. Underglaze applications are made permanent only by the layer of glaze over them, which protects them from the ravages of wear. Obviously, for these colors to be visible, transparent or translucent glazes should be used as a cover. Underglaze pigments are suitable for application on unfired glazes, as in majolica.

Overglazes are formulated for application on fired glaze surfaces. The compound includes a flux, generally in the form of a frit. After the overglaze decoration has been put on the pot, it is fired once again. This third firing is called the *decorating fire* and takes place at a low temperature, generally in the neighborhood of cones 015–012. The decorating fire melts the flux or frit in the overglaze, fusing it to the glaze. In that sense, overglaze is somewhat like a very low-fire glaze layer. Overglaze decorations do not easily wash or wear away, but the coating is not as resistant as underglaze to the friction of wear. It is often possible to detect an overglaze application by close examination. It appears as a distinct layer, and its texture seems subtly different from that of the glaze surface underneath. Because they are fired at considerably lower temperatures, overglazes are available in more vivid intensities than underglaze colors.

As might be expected, underglaze and overglaze pigments may be purchased in quantities convenient for the studio potter. They come as powder, and also in watercolor sets, crayons, pencils, in liquid form, and also as "magic markers". In these preparations, they are particularly attractive for use by children.

Potters experience color differently from painters. For the potter, color does not come out of a tube, its nature immediately revealed and faithful to the label on the tube. Color manifests itself eventually, in reaction to a series of pottery techniques and processes. The potter is very apt to think in terms of oxides: cobalt, for example, rather than blue or copper instead of green—the productive oxides in creative thought and action—rather than the names of colors. Learning about color in pottery means understanding how colorants react to materials and fire, application techniques, and the forms in which the pigments are used. These merge to become the potter's creative instrument.

The most direct and certainly the least expensive way to introduce color in the pot is to use raw metallic oxides, rather than the commercially prepared stains described above. Of the two, some potters think of the oxides as the more intimately linked to the materials of the earth and, therefore, philosophically closer to the family of clay.

However, colorant oxides are generally not as uniform or predictable as the commercial stains. While this state of uncertainty may appear at first to be a disadvantage, many potters are invigorated and challenged to work with attractive natural materials. They furnish the potter the potential for a very extensive palette. The oxides are easy to incorporate in all the stages of pottery making: slip and body preparation, application on unfired and fired wares, putting in and on glazes. They result in color qualities that can reach out naturally from the pot and that may in execution become visually more varied and intriguing. In many studios and schools only the oxides are available for color use, and much good work results. But from time to time, a particular oxide may become scarce and expensive. The creative and resourceful potter takes these developments in stride and finds suitable alternatives. Color as decoration is merely one of many elements on the pot, and it is often not the most effective one.

There is no "best" way to apply the different varieties of colorants, whether raw metallic oxide or prepared stain. A suitable *vehicle* must first be mixed with the dry powder. (The vehicle is the liquid that conveys the pigment.) Water is the most convenient vehicle and the most often used. For certain special effects and techniques, fine oils, such as oil of lavender, are more suitable as vehicles.

Brushes are probably the most traditional and most utilized tools to stroke and paint color onto pottery surfaces. There are pointed and flat, long-haired and short-haired brushes, and many shapes and sizes to choose from to make lines of great variety and rich quality. Each brush shapes a line of distinctive character. Short, stiff, bristled brushes may be used to stipple color onto the piece, resulting in groups of points, resembling mottling.

The potter may also choose to air brush or spray color. Stencils may be used to block out and define sharply the flat shapes and patterns of sprayed mist. A very direct approach is to pick up some of the color liquid in a coarse, dry sponge, and pat or stamp it on the pot's surface, leaving a characteristic clouded and rippled surface. Colorant may be spattered over the surface by rubbing a color-laden toothbrush across a stick, close to the pot. Splash and pour and dribble color. There are no fast rules.

Each application method has a unique personality, which becomes one of the components of composition arranged on its surfaces. Choose an application method in character with and compatible with the design of the whole decoration and

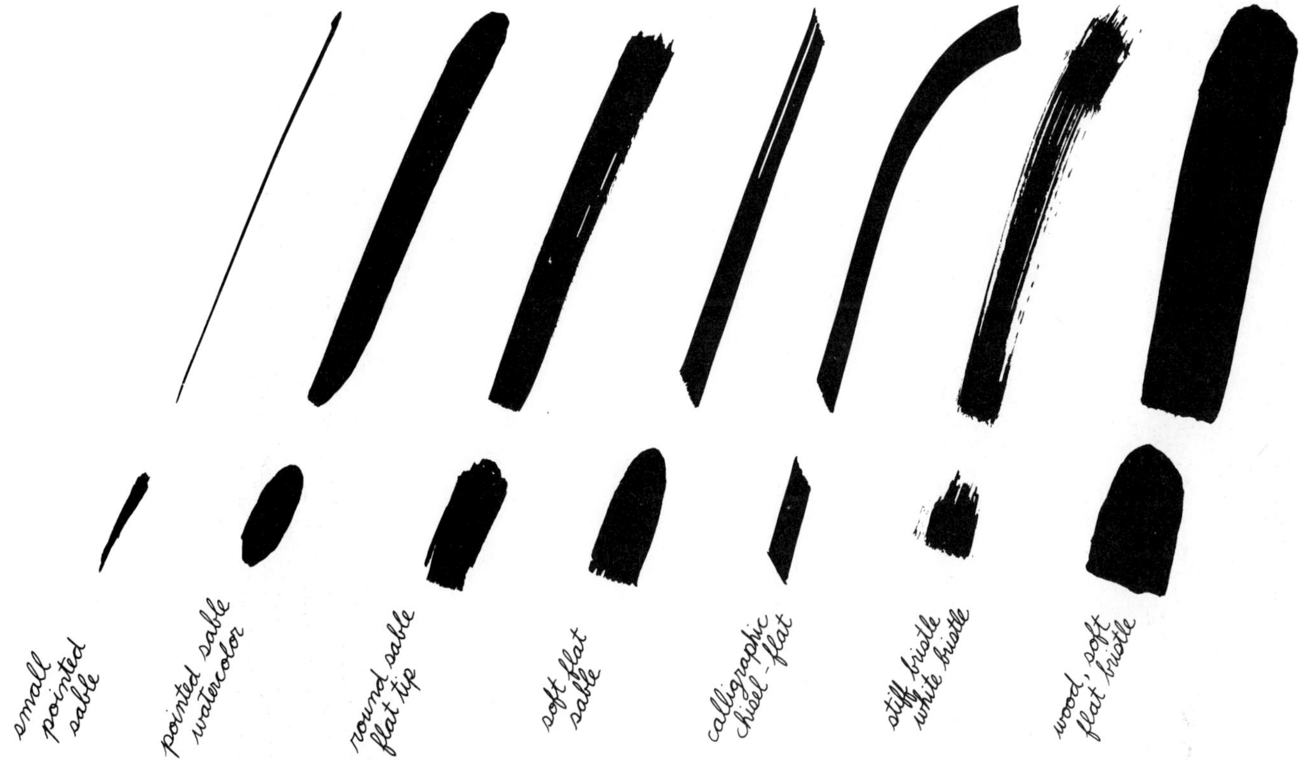

9-2 *A great variety of brush strokes are possible.*

9-3 *Color sprayed through a stencil. Note the sprayed texture.*

9-4 *Color stamped with a sponge.*

the pottery form. Graceful, brushed calligraphy may prove unsuitable for some pots even though, viewed separately, the brush stroke may be breathtaking. The dominant purpose of decoration is to enhance and reinforce the pot's form, and that is the principle measure of its success.

Texture

Texture is the surface quality that stimulates the sense of touch. We call that the tactile sense. The potter shapes clay by the pressure of hands and squeezing fingers. Therefore, it is not surprising that the tactile appeal of clay is one of its strongest components. We sense textures through the nerve ends at the skin, particularly at the fingers and the palms.

But we also appreciate texture visually; we see textures even when we cannot feel them. Sometimes we see textures that really are not there, as in photographic representations. The sandy beach evokes sensations of the graininess and soft sediment, even though when rubbing the fingers across the photograph it feels glossy and smooth. We see surface qualities of any kind by virtue of the changes in light and dark in those surfaces. From smooth to grainy to coarse to pitted, each visual variation is some pattern of light and shadow.

A pattern is an orderly array in harmonious repetition. When a potter deliberately evokes tactile

9-5 *Textures made in plastic clay with fingers and tools.*

9-6 *Variations in throwing marks.*

responses to his work, he inevitably arranges changes in depth at the clay's surface with fingers and tools. Ridges, valleys, scratches, crumbles, pits, marks, points, and scrawls—all these textures are visible because of how they capture shadow in the recesses and reflect light at the crests. Generally, the more interesting textures are repetitive; the eye can then more intelligently take in the meaning and the impact. A sensitive potter often generates interplay among several textures, rather than covering the surface with uniformity.

Probably the most integrated textural decorations are those which arise from the clay's properties, shaping techniques, and tools. Clays and pottery bodies are often formulated to focus on certain textural effects. Some kinds of pots seem more comfortable made of clay with a pronounced "tooth", while others work best with fine grained smoothness. The simplest way to change the texture of a clay from smooth to coarse is to knead into it a measure of grog, of the grain size desired.

Throwing clay on the potter's wheel produces a fluid, spiral groove which covers the entire surface. The potter's fingers develop a line of light and shadow which encircles the piece, moving sequentially from the bottom to the top, a beautiful complement to the clay's plasticity. With more experience and growing sensitivity and awareness, the potter can more skillfully modulate the thrown textural pattern. Press harder with the finger tips, turn the fingers for blunt or sharp contact, use a different finger tip, turn the wheel a bit faster, change the throwing pace. Observe how the depth and spacing of ridges and valleys can be arranged to accentuate flow in the pot's volume, such as a fat belly diminishing into a slender neck.

In pinch pots and coil-built forms, pinching, lovingly executed, leaves a patina of small, pebble-like depressions, a beautiful expression of volume surrounded by surface and at the same time a wonderful ground for glaze. The pinch marks may be made dynamically more interesting by varying their intensities. Pinch harder to make the depressions deeper, pinch closely spaced or further apart. The visual pattern of light and dark follows in kind.

Trimming tools used to cut the footrings and footings of thrown pots also leave a continuous ridge at the bottom of the wall. Unlike the soft, fluid, plastic movement of finger marks, trim lines are hard and mechanical. Potters are aware of the need to integrate tool cuts with finger marks for the two qualities to be in composed adjustment.

Paddles beat pots into shape, and as a by-product leave on the clay surface a faceted network of flat planes. The sharp edges of scrapers used to refine handbuilt pieces into more sophisticated silhouettes result in a smoothness of uninterrupted contour but a tactile grittiness. All these textures derived from shaping methods and tools may be sensed by touch, but they are also intended to be seen as an element of the pot's visual organization.

In good work, these visual surface manifestations are designed and executed to complement the significance of three-dimensional form.

Clay is the easiest material in which to introduce textures. Any tool or instrument, including finger tips and nails, may be used for tactile effect. Impress foreign materials, such as string, rice grains, wood shavings, and grasses, into the soft surface. Stamp the ends of broken sticks, jagged rock, crumpled newspaper, dry sponges into the plastic clay.

Line

A mathematical line may be the shortest distance between two points, but a line in decoration is endowed with much more significance. Lines are what we see in the pot's silhouette: outlines. Lines have direction; they cause our eyes to move from one location to another. There is movement and rhythm in lines, straight lines giving off a different energy from that of curved lines. A sinuously curved line has a rhythmic quality very different from the nervousness of a jagged, broken line. Lines tell us about power and inertia in the piece. Vertical lines are stable and strong; horizontal lines show a restful condition; and diagonals or obliques are energetic and active.

Lines are visual movement of one kind or another. The line does not actually move, but the spectator's eye is conducted along a pathway. As the eye is directed by the line, it experiences within the line still another dimension, that which we call *line quality*, whether the line is uniform or varied, thick or thin, precise or soft, pointed or blunt, solid or translucent, flat or tapered. An unlimited source of rich material for the potter to enrich clay surfaces is in line quality. Each variety of line is a distinctive spirit in the life of decoration. Rich line quality is much more interesting than mechanically uniform line.

In their rhythms and qualities, lines function in the most fundamental sense as abstract visual elements. But lines may also be symbolic, by telling stories. Lines may be drawn on pots to represent natural and other categories of objects—people, fish, or trees—and to reflect the potter's inventions and fantasies.

Lines may be scratched, drawn, and painted on pots at any stage of the pottery sequence. Scratching or incising in soft clay sometimes leaves a burr along the edge of the incision. These raised edges and crumbles are best smoothed and scraped clean when the piece is dry, but care should be taken not to lose the crispness and freshness of the original drawing. It is prudent first to practice on some scrap clay, for it will be difficult to erase mistakes on the good pot.

Brushes are often used to paint colored lines with oxides, engobes, underglazes, overglazes, glazes, and wax. Breathtaking examples of brushed calligraphy may be seen in traditional Chinese and Japanese pottery. In these, line quality is exercised sparely in some pieces and with lavish expression in others, to the end that these lines capture the fullness of human life and spirit. Much of this timeless work is unsigned, the work of anonymous craftspersons. The Eastern legacy of brush calligraphy is very influential in much contemporary

9-7 *Variations in line quality.*

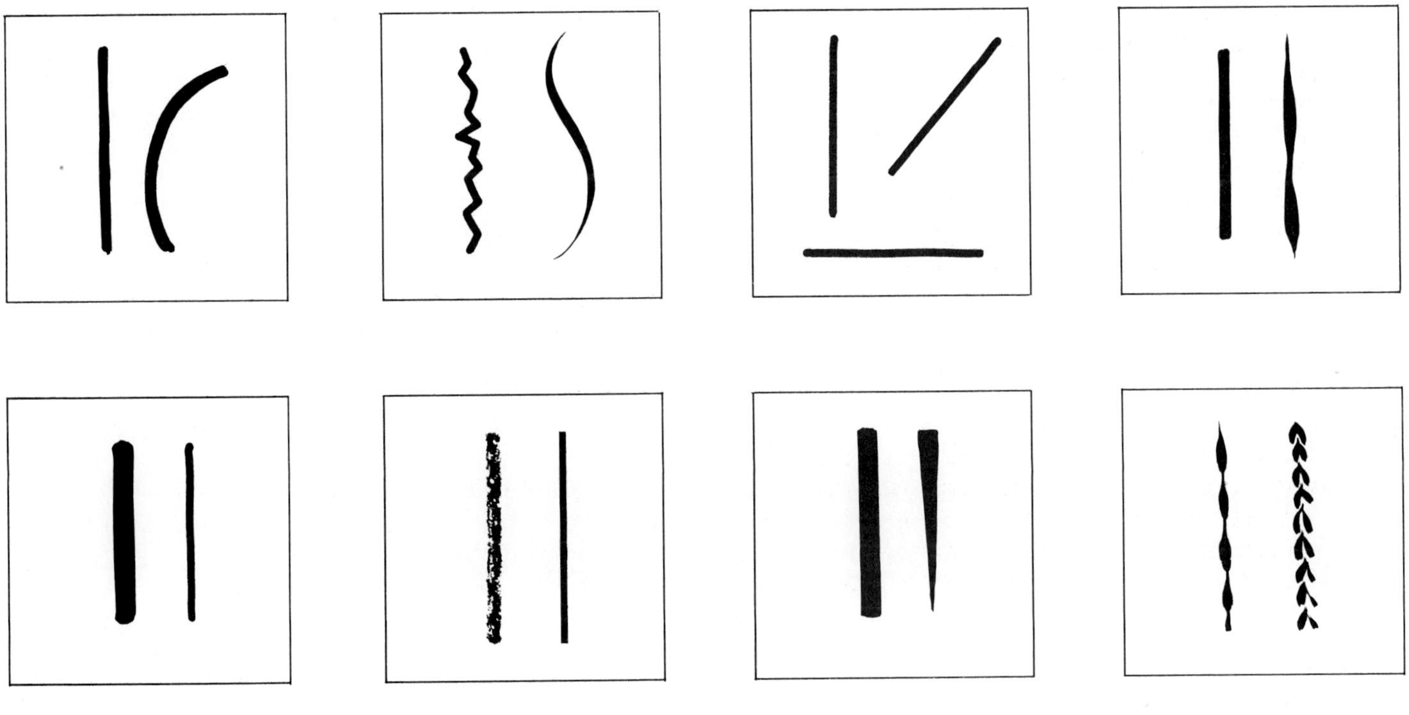

9-8 *Lines scratched, drawn, and painted.*

pottery. However, to the Western hand the use of a brush does not come as easily. Much practice of brush strokes and composition, first on paper and then on waste clay slabs, should precede painting with the brush on good pots.

Shapes

In this book we have expressed the word *form* in a manner to distinguish it from a similar concept, *shape. Form* encompasses the pot's volume and in some instances its mass, the piece as a vessel or solid in three-dimensional space. *Shape* is a flat configuration on the pottery surface. It is generally two dimensional, lacking any significant depth. Some shapes are made more distinct by an outline drawn around them. Other shapes are distinguished from their surroundings solely because they contrast in color, texture, or value.

Like lines, shapes also have abstract visual meanings. It is easier to perceive basic abstract elements in color than in lines and shapes. Color has an obvious direct appeal to the senses, such as one's favorite color. Lines and shapes frequently take on other associations and meanings, such as representations of recognizable objects. These may get in the way of understanding and appreciating the essential traits of lines and shapes.

Shapes may be arranged to give the impression of direction and movement covering the surface of the pot. By means of configurations, repetitions, patterns, and overlappings, diverse rhythmic impulses may be generated. Shapes may also be com-

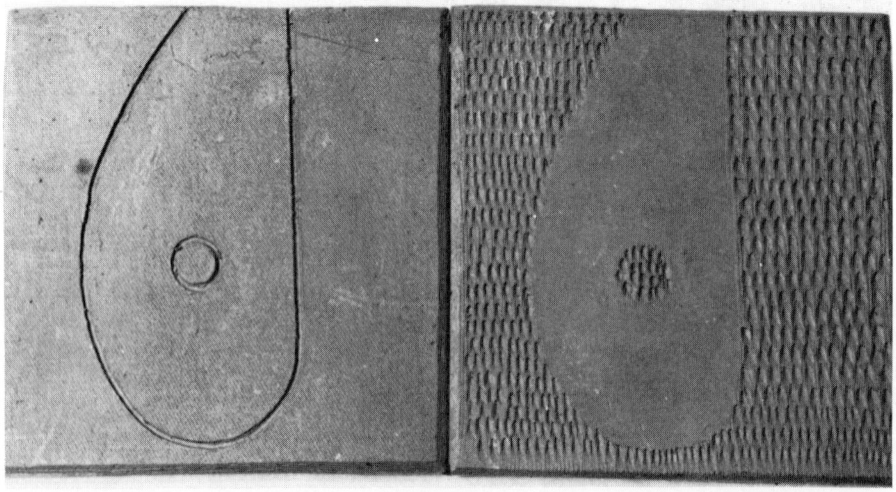

9-9 *Shape may be defined by line (left) or created by contrast (right).*

posed to suggest moods and feelings; they could be languid, droopy, runny, jagged, stolid, staccato, graceful, elegant, architectural, elusive. Placed strategically on the pottery surface, they serve as focal points for the eyes to rest and to engage the eyes in the business within.

Shapes may be placed primarily to assist us in grasping more completely the full meaning of the pottery form; shapes in decoration, largely for the manner in which their configurations and locations are created, strengthen the form. But shapes on a pot may also represent additional perceptions, the pot serving as a hospitable environment for tales to be told. Stories and fables unfold—ancient Greek pottery furnishes many rich examples of recorded history on pots. Objects from nature are fertile materials for shapes in decoration, or shapes may be invented from the potter's unfettered imagination.

VISUAL ELEMENTS AND POTTERY FORM

Some pottery is exquisite because of its coloration. On some the calligraphy seems to epitomize the height of human sensibility; on others we are overwhelmed by expressive imagery. But whether the decoration is ordinary or succeeds to the sublime, the potter is always confronted with the serious problems of whether the decoration suits the pot and how to place it for most comprehensive meaning. The reader might be helped in executing well-composed and appropriate decorations by the following precepts:

1) Does the piece *need* any extra embellishment or decoration other than a glaze? Sometimes the potter only wants to display virtuosity and extrordinary skill in some decorative technique. Or perhaps the decoration originates as a cover up or camouflage of weak form or poor shaping craftsmanship. If in doubt, it is better to leave it out.

2) While total consistency between surface and form may be totally boring, the lack of integration between them may result in visual chaos and anarchy. We can call that condition *the destruction of the pot's form*—figuratively, not literally. Does the decorative style suit the form, or do the two clash? Is the decoration appropriate and compatible in its abstract essentials and symbolic elements?

3) Generally, a useful guide is that in vigor of line or shape and strength of color, the surface decoration should neither dominate the pottery form nor overwhelm it. But there are too many exceptions in which a pot's existence is justified primarily because a miraculous expression rests on its surface.

DECORATIVE TECHNIQUES

Materials, Tools, and Forming

Probably the most natural, and certainly the least self-conscious methods for embellishing pots originate in the clay's characteristics and how clay is shaped. Smooth grained clays have an intrinsic slickness, which can be altered by mixing or kneading the clay with assorted textural sources, such as sawdust, rice grains, or grog. Except for grog, these materials are combustible and burn away in the firing. The sand-like grains of grog remain permanently within the clay's structure. The texture of grog is apparent even under a glaze. These ingredients reduce plasticity, proportional to the amount added to a clay. It is an important consideration when high workability is needed in a pottery body, as for throwing. (A full discussion of clay and body formulation and alteration is in Chapters 4 and 5.)

Pinching, coil building, slab building, and throwing each leave in the pottery surface characteristic evidence of which of these shaping techniques has been used. Pinching results in a patina of small impressions, the pinch marks left by finger tips. In coil building, the clay walls are often also pinched, but, if scraped or paddled, these tools leave their mark in the surface. Coils may be left to show as tiers, strong ridges or subliminal reminiscences of how the pot was structured. Slab building is characterized by large, open faces which tempt the potter to leave a varied supply of marks on them. Slab faces may be made richer merely by altering the clay's texture and modifying the surface with tools and color. The sign of throwing is a continuous spiral climbing up the wall made by finger tips squeezing on revolving, soft clay.

The sensitive potter composes and arranges these characteristic shaping impressions, patterns, and textures for the good of the pot's form. Marks may be set closer together or spaced more openly. They may be patterned uniformly, or scattered erratically. Impressions and marks may be made deep or shallow, for weaker or stronger light and shadow contrasts. Morever, since pots are customarily glazed, the potter applies his shaping skills consonant with the traits of the melted glaze. Some glazes pick up surface inequalities and enhance them, while other glazes may tend to obscure them or even cover them completely.

Shaping tools and instruments may generally be counted on to erase finger marks and substitute their own idiosyncracies. Paddles and scrapers used in handbuilding smooth over pinch marks and coil layers; facets and smoothed contours appear instead. Ribs used in throwing give the pottery form a more swiftly flowing surface, perhaps a tighter and more controlled outline. Trimming tools for footrings and the like produce spiral ridges, but unlike those from throwing finger tips; tool marks are harder and sharper in their light and shadow rhythms. The surface effects used to shape pots are as much elements in decoration as any more consciously and deliberately applied technique, such as incising or underglaze painting. Therefore, the reader should always be alert to the impact of shaping methods on surface quality.

Incising

Scratch or draw into soft or leather-hard clay with a pointed tool. A chisel-shaped point works better than a round point. The temptation to use a pencil or a needle is really misplaced. It creates a chicken-scratched, uniform line, uninteresting and unexplored in its potential. The round point tends to leave a messy burr along the length of its edges.

Bend the tool one direction or another while drawing, to give the line variation in thickness. Line quality can be further enhanced by drawing more deeply or with greater subtlety. Lines drawn with assurance contribute more to successful decoration than wavery scrawls. Practice on scrap clay slabs. Incising works best in leather-hard rather than soft or bone-dry clay. A burr along the edges of the incising sometimes occurs. If left, the raised crumble would be unsightly under the glaze and would feel like a harsh welt. Remove the burr by careful scraping when the piece has become bone dry, but with care to retain the clean and fresh spirit of the drawing.

Carving and Modelling

These are opposites in the sense that carving is executed into the pottery surface to create an intaglio (negative) shape or plane below the surface level, while modelling means adding to the pot's surface, resulting in built-up shapes, planes, and reliefs. The effect of modelling is seen on a coin or medallion. Carving is excavation. Work when the clay is leather hard or softer. For modelling, weld bits of clay, small coils, and thin slabs into place with just enough moisture to make them stick. Differently shaped modelling tools are effective to strike off the contours in negative and positive relief. For extremely fine work specially shaped steel tools are available.

Stamping

Press any object that will leave its mark into the plastic clay surfaces. Scrap materials do well: hard, crumpled newspaper; broken chunks of plaster; jagged ends of sticks; small stones; coiled string; wound rope. Press implements from around the studio into the clay: wire loops of trimming tools, tongue depressers, modelling tool and potter's knife ends, type letters and numbers, finger nails, screw drivers, hammer claws. Stamps can be made solely for stamping, and butter and cookie molds may be

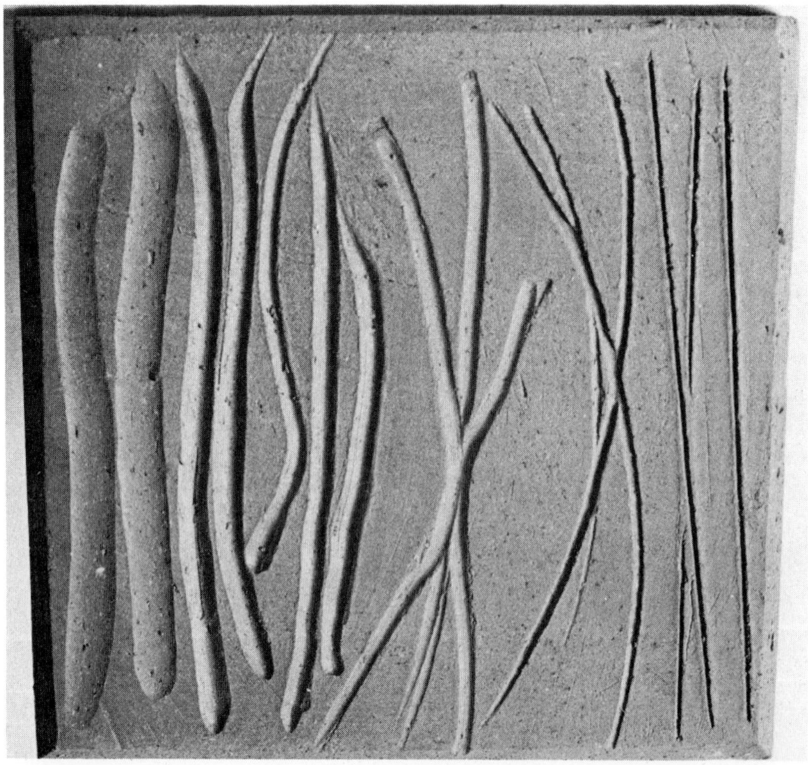

9-10 *Variations in incised lines.*

9-11 *Carving and modelling.*

9-12 *Stamping.*

used. The shapes may be abstract, geometric, or figurative. Stamps are easily made of clay, to be used after the bisk fire, or carved by a knife into plaster or wood. Some stamps are shaped into a wheel to which a handle is attached. These "roulettes" press a continuous sequence of the mark when rolled across the pot.

Incising, carving, modelling, and stamping are decorative techniques which yield differentiated levels at the pot's surface, either built up or taken away. Semi-opaque glazes display them to their best advantage. When the glaze melts over the pot, it pools in the recesses and thins over the convex surfaces. The glaze appears more transparent where it is thinnest, and more opaque where it is thickest. These gradations and modulations of translucency accentuate the decoration.

Slip Painting and Trailing

Decorative slips or engobes are most often applied when the pot is leather hard or softer. In this way the slip, which is essentially clay and water, will most successfully make a lasting bond with the pot. The fundamental character of slip is its dimensional thickness, resembling artist's oil paint, and its opacity or ability to hide what is underneath. Therefore, slip makes its appearance as a generous layer over the pot's surface. Use a soft bristled brush, pointed or flat, to apply the slip. The brush stroke in its textural variations is readily apparent in the slip. Where the brush's bristles have left the slip thin and drawn out, the slip is likely to become fugitive—it disappears under the glaze, having been taken into solution by the melting glass. These subtle variations are attractive, and they seem to unite the decorative line of slip with the substance of the clay. It is prudent to practice first on scrap clay, for erasure is difficult. If mistakes are made, the whole decoration must be scraped off clean, unavoidably removing finger marks as well as other attractive textures.

For *slip trailing,* fill a plastic squeeze bottle, such as a ketchup dispenser or ear syringe, with slip. Draw the slip into the empty container smoothly all at once, to avoid sucking in air bubbles trapped in the thick liquid. Move the trailer's point over the pottery surface, squeezing steadily to extrude a trail behind the moving point. One works quickly, without pause or hesitation. Slip trails are welts, raised lines, squiggles, drools. There is a resemblance to squeezing toothpaste or cake decorating. The line forms quickly; it cannot be studied. If errors are committed, the best remedy is to erase by scraping away the trailed line, but only after it has dried firmly. Beautiful examples of slip trailing may be seen in Colonial and early Pennsylvania Dutch pottery. (In Chapter 8 there is a thorough description of engobes and color.)

Sgraffito

The word comes from the Italian, meaning "to scratch through". Apply a coating of engobe over an area of the pot by brushing, spraying, or dipping. The engobe color should be different from that of the clay. It may be put on as a uniform, opaque coating or a variegated, partially fugitive layer. When the engobe has dried to at least leather hard, incise through the slip layer to the pottery surface underneath. The drawing is exposed by the contrast in color between the exposed line and the engobe. Line quality is a critical element in the effectiveness of this decorative technique. Sgraffito permits the potter to engage in some very fine and sensitive line expression.

9-13 *Slip painting.*

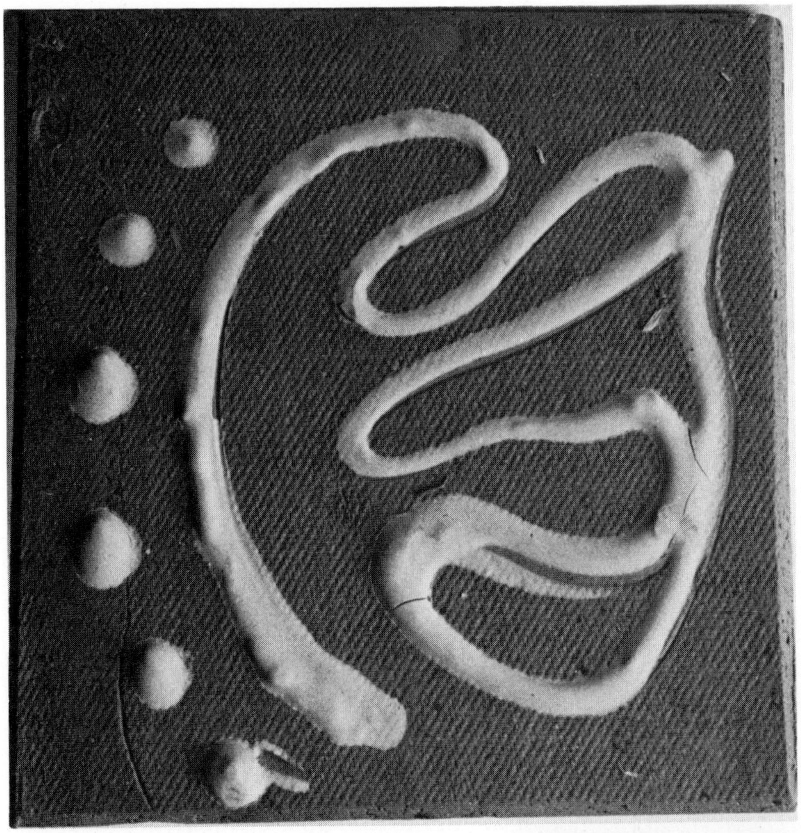

9-14 *Slip trailing.*

9-15 *Sgraffito.*

Inlay

An incised or stamped depression in the pot's surface is filled flush to the surface level with engobe. First incise or stamp into plastic or leather-hard clay the desired composition. Inlay invites the application of single decorative units, particularly pattern and repeated motifs. With a flat brush, paint over the impressed section until it is more than covered with slip. After the pot has turned drier than leather hard, scrape away the slip down flat to the surface level of the pot, until the deco-

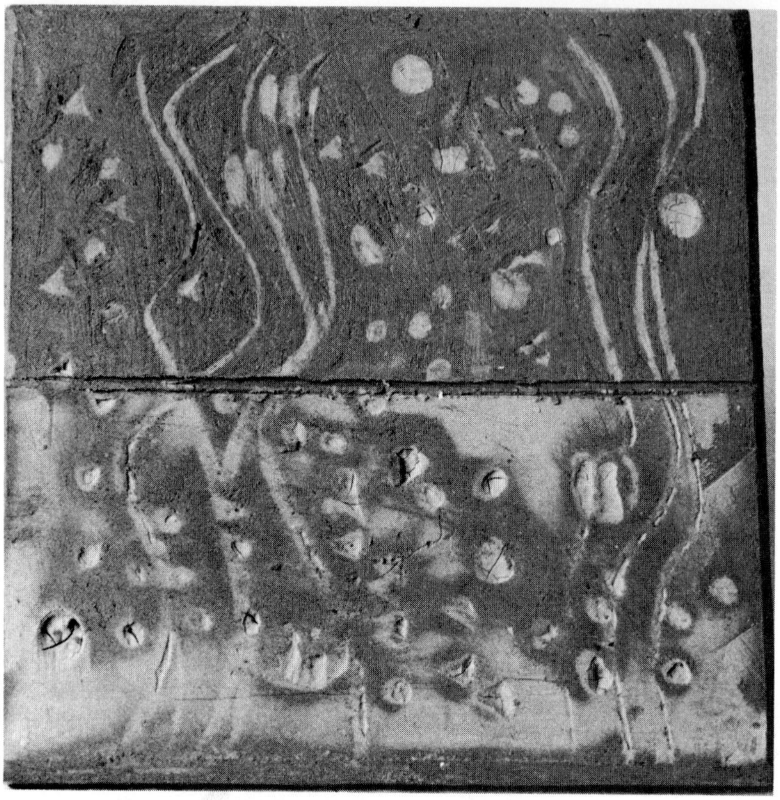

9-16 *Inlay (top) and mishima (bottom).*

rative units are cleanly and crisply revealed. Slip remains in the depressions, but the surface is bare. Carefully scrape at the surface until the drawing or stamping is shown flush and precise. Use a slip color which contrasts with that of the pot, such as white slip on red clay.

In Korean and Japanese pottery there is a similar decoration, called *mishima*. The technique as practiced by these potters of the Far East is derived from a view of existence which differs from the Western pottery tradition. They do not scrape clean around the incised or stamped decoration. Some of the slip is left as a casual residue around the intaglio. This effect reflects, in their view, continuity, integration, and linkage the same as in all of life.

Slip painting, sgraffito, and inlaying require the application of transparent or translucent glazes over them. Opaque glazes would cover and hide the effort.

Wax Resist

This mode of decoration permits layers of lines, colors, and shapes to be applied, in turn, on top of each other, but each incremental layer reveals distinctly a portion of the preceding one. The theory that makes it work is that liquid wax applied at any stage of the pottery process repels or resists watery mixtures, such as engobes and glazes. Any of these liquid coatings poured, sprayed, or dipped over a waxed area simply runs off the wax. In the firing the wax vaporizes and disappears. Paraffin may be used alone, but the method is more effective when the paraffin is mixed approximately half and half with beeswax. An imitation beeswax is available that is much less expensive. The best way to prepare the wax mixture is to melt it in a double boiler or in an electric fry pan, where a low temperature can be maintained. For safety's sake do not melt paraffin directly over a flame or hotplate. The resist mixture must be kept warm and melted during use. It can easily be remelted repeatedly. Potters may also purchase an alternative resist material, a liquid-wax emulsion very much like liquid latex. It does not require heating or melting—its major advantage. Unlike wax, this liquid resist can be washed from brushes with warm water and a detergent.

Most potters like to apply wax resist with a brush. Parts of pots can also be coated by dipping them into a melted wax bath. The wax hardens almost immediately. Lines and patterns may then be scratched through, similar to the batik method for decorating cloth. Wherever there is some wax on the clay, that section will resist whatever coating is applied. Sometimes, however, a few drops or drools of slip or glaze, rather than sliding away, gather here and there on the wax. These drops may be easily removed by blowing them away while wet or wiped aside with an elephant-ear sponge.

Complex decorative compositions are easily ar-

9-17 ***Brushed wax resists glaze.***

ranged by using wax resist. For example, several layers of slip may be interspersed by strokes of wax. Following bisk firing, more wax can be brushed to interact visually with the slip layer underneath, after which a glaze is applied and more wax pattern brushed on the unfired glaze, to be covered in turn by a second glaze layer or a colorant. At each wax application the potter may scratch through lines, dots, and textures. However, wax resist is often used very simply at only one stage, and with very interesting effects.

Errors committed in the application of the wax are difficult to erase. On leather-hard or bone-dry pieces, the cooled wax can be scraped away down to the level of the clay. Unfortunately, some damage to the freshness and crispness of form and surface may result. On bisk pottery it is best to fire the pot once again for the wax to burn off completely. Wax applications on unfired glaze have to be scraped away, unfortunately together with all the glaze. Therefore, it is prudent to plan ahead, carefully and precisely, where the wax is to cover, and mark appropriate guide lines. Practice first with the brush, for deft, swift, faultless application.

Somewhat similar results can be obtained with masking tape or rubber cement on bisk wares or fired glazes. Engobes and glazes may then be applied over these masking materials, which burn away during the firing. It is an alternative but not an equivalent approach to the use of real wax resist.

9-18 ***Wax resist decoration after the glaze firing.***

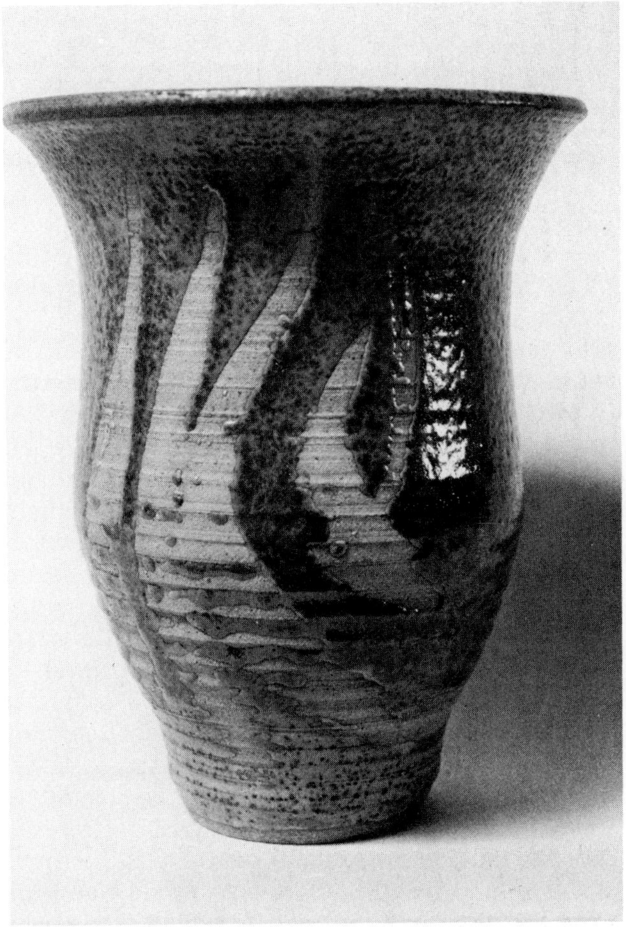

10 DRYING CLAY WARES

Clay wares must be completely dry before being placed in a kiln for bisk firing. Kilns are poor dryers, totally enclosed as they are for heat retention. The advance of heat in the early firing stage would be rapid enough to harm moist pots by causing the clay to crack or explode. Clay rarely blows up at any other time during firing. Were moisture left within the clay's pores, the rapid advance of temperature would turn the water into steam, which, finding its way through the clay structure, would force it to explode. The desired condition of complete dryness, wherein all but atmospheric moisture has left the clay, is called *bone dry.* Bone dry can be determined by touching the clay and by observing the color of the dry pot: The clay should feel near to room temperature, not cool or clammy and should look chalky.

Unavoidably, whatever humidity is present in the air in the studio is also present in the pores of the dry pot. This residual moisture is sometimes called *pore water.* During the early, slow buildup of several hundred degrees in bisk firing, this type of water, intimately intertwined within the clay compound, is gradually released in the form of steam. The vapor leaves the kiln through peepholes, cracks, openings, and chinks.

All clays shrink during drying and firing. During drying, this shrinkage takes place when the clay particles draw closer together as moisture departs. While shrinkage is inevitable during drying, it is not desirable that the loss in size be excessive or uncontrolled. Above 15% total shrinkage might be considered excessive. Control of drying can be accomplished by the composition of the pottery body, by additions to a single clay, and a favorable environment arranged for drying wares. (In Chapter 4 there is a complete explanation of clay and body formulation.)

The environment surrounding the drying pot is of critical importance. In pottery factories specially engineered dryers control with ideal precision such factors as air circulation, temperature, and humidity around the wares. These dryers are frequently heated by hot waste gases drawn from the kilns and by infra-red lamps. In this manner, industrial clay products are brought to a bone-dry state properly, safely, and rapidly. Unfortunately, this advanced level of technology is rarely practical for an artist's studio or a small pottery.

Pots come in a variety of forms, sizes, and wall thicknesses. The most practical method to dry this assortment is out in the open on shelving. The shelves should not be close to direct and strong heat or draft. While metal shelving is usually readily available, wooden shelving is preferable. It is porous, non-rusting, and soft enough not to mar or dent still-soft foot rings. It is even more efficient when constructed of parallel slats spaced about 1/2″ to 3/4″ apart, which permits air circulation from underneath. The wood should be left unpainted and rigid enough to support without warping large and heavy pieces. Artificial dryers in enclosed cabinets and heat lamps focused on pottery out in the open are to be used, if at all, with extreme caution. The danger from these devices is warping and cracking from uneven, rapidly rising, and excessively concentrated heat.

Special attention should be given to pots with

handles and lids. Lids should, whenever possible, be dried and fired in place on their pots. They settle down together and in that position are mutually supportive. Pots with freshly attached handles should be first set in a damp cabinet or covered with plastic, at least overnight. This gives the clays an opportunity to come together and adjust, creating the strongest bond at the joints. Afterwards, on the drying shelf, a slightly damp cloth or strip of plastic loosely draped over the handle part helps equalize drying speed. The idea is to prevent the handle from drying faster than the pot, minimizing cracking at the joint from uneven shrinkage.

Generally, it is best to dry pots on their feet. In this position they support themselves more substantially. Pieces with very wide rim diameters, such as large platters and bowls, might, at intervals, be turned over to dry on their rims. When pots are placed on their rims for drying, first place a layer of material on the shelf to permit the piece to breathe from underneath and to shrink freely and without obstruction. A sheet of newspaper or a light dusting of potter's flint, sand, or fine grog on the shelf permits the pot to shrink easily. In this manner warping can be minimized.

It is difficult to generalize about the length of time it takes for a pot to become bone dry. Variables in body composition as well as studio conditions are complex influences. Thinner and smaller pieces dry faster than thicker and larger ones, coarse grained clays more rapidly than finer grained bodies. It is safe to say, "In a day or two."

11 TYPES OF KILNS

The effect of heat on clays and glazes is central to the art and craft of pottery. Kilns are the "ovens" in which heat is generated for firing to take place. Potters have always designed and constructed many types of kilns with great ingenuity. In early times, however, there were no kilns as we know them now. Probably, the first firings took place in open fires, the same source as that for cooking, heat, and light. Soon, in separate firings, the dry wares were just piled up with a wood fire laid around the heap of pots. There are some places today where this relatively crude method of firing is still practiced. Soon after, firing was improved to the extent that it was done in a hole or pit dug in the ground, with a fire built above and the red embers sifting down through the pots. In recent years there has been a renewed interest in this primitive method. Then came firing in caves, to be followed by the prototypes of today's kilns, simple structures built of stone and wattle. Finally, planned and designed kilns were made from manufactured clay bricks. Modern day kilns are fabricated from standard shapes and sizes of special refractory ceramic materials, bricks, sheets, slabs, and insulating blankets.

The essential components of a kiln are the refractory structural materials with good insulation properties, a means of generating and controlling heat, and an escape route for gases. There is a chamber in which pots can be stacked safely and efficiently during the firing. In the past, all kilns were heated by burning natural fuels: dried animal wastes, peat, coal, wood, oil, and gas. Although all these fuels are in use somewhere today, most studio combustion kilns burn gas or wood.

A fairly recent development in the history of kilns is the invention of the electric kiln. No fuel is burned, but heat is generated by the passage through a resistance of electric current, similar to what takes place in toasters and electric ranges. The electric kiln is essentially different in its effect from combustion kilns, in that the air or atmosphere inside the kiln is not an active participant in the firing process, while in combustion kilns the nature of air is a basic component. Electric kilns are widely used today for their convenience and safety.

The modern kiln, whether combustion or electric, is undoubtedly an efficient instrument compared to that of the past. Newer structural materials are remarkably efficient heat insulators; heat resources may be consumed more economically, and the generation and control of heat is much more precise. Kilns are now available in which extremely high temperatures are realized more easily. Kiln atmosphere has become for the modern potter a more reliable firing tool. There are small as well as large kilns, suitable for the diverse needs of part-time potters, hobbyists, schools, craft centers, and full-time studio potters, and there are accessories available which make firing more effective or at least simpler.

COMBUSTION KILNS

The two basic construction styles are the *open kiln* and the *muffle kiln*. The fuel chosen for the kiln is the determining factor for special design arrangements to burn and control that fuel, as well as how to provide for draft and exhaust for waste

11-1 *Pots in an open fire.*

11-2 *Pots in a pit fire.*

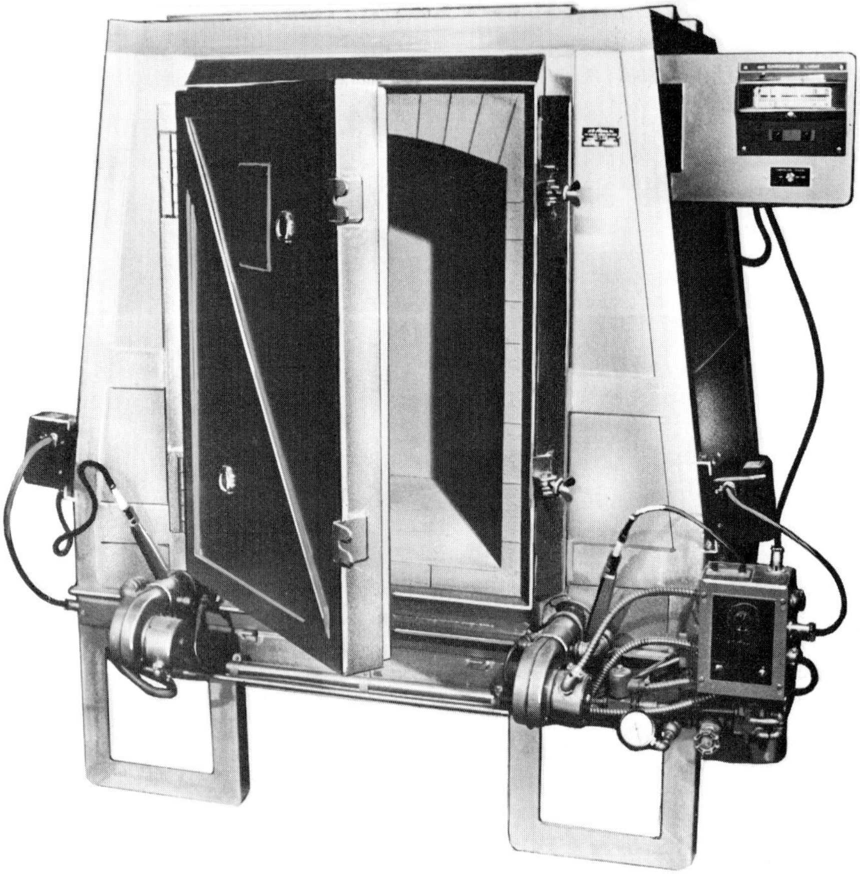

11-3 *A studio gas kiln.*

A. D. Alpine, Inc., Torrance, CA.

gases. Gas kilns may be used indoors, while wood-burning kilns may not. There are mutations, adaptations, and compromises in design, but the two basic types may be clearly understood as follows.

Open Kilns

This design is called "open" because the chamber in which the pots are stacked is the same in which the fuel is burned. Inside the kiln, the wares are open to fire and flame. Color and textural variations in the melted glaze often result from the flashing and licking flame. In wood-burning open kilns, ash residue from the fire often falls on the pots, creating interesting ash effects. Most studio potters welcome these erratic by-products for their rich array of visual and textural variations. No matter how intelligently designed the kiln or skillful the potter, there always seems to be at least some room for the chance occurrence—changes in color quality or in the texture of the glaze or the exposed body surface, sometimes ever so subtle and subdued, at other times strong and declarative. The piece seems transformed, brought vibrantly into a new life.

However, some pottery wares, especially products of standard grade such as tableware and other industrial pottery, require flawless and uniform glazes. If a muffle kiln is not available, then the glazed work must be stacked in *saggers* for the glaze firing. Saggers are large containers made of special refractory clays. One or more pots are placed in a sagger. The saggers are stacked in vertical rows in the kiln. Only the top one needs a cover to protect the wares inside from open flame, ash, and erratic atmosphere.

Open Updraft Kiln This is a variation of the open kiln in which the chimney or flue opening is through the crown or ceiling and the heat source, wood or gas burners, is at the bottom. The direction of the fire, heat, and gases of combustion is up, vertically towards and out through the chimney. The action of the heat on the wares is, consequently, somewhat inefficient, as the passage of heat is swift and direct. In comparison with a downdraft kiln, it is more difficult to maintain uniform heat throughout the firing chamber; there is often quite a variation in the heat obtained in different parts of the kiln. For example, the bottom back might be 50° or so hotter than the top front. Kiln atmosphere fluctuates more in an updraft kiln. These variations may be of little concern in bisk firing, but they surely will leave telling marks on glazes. The

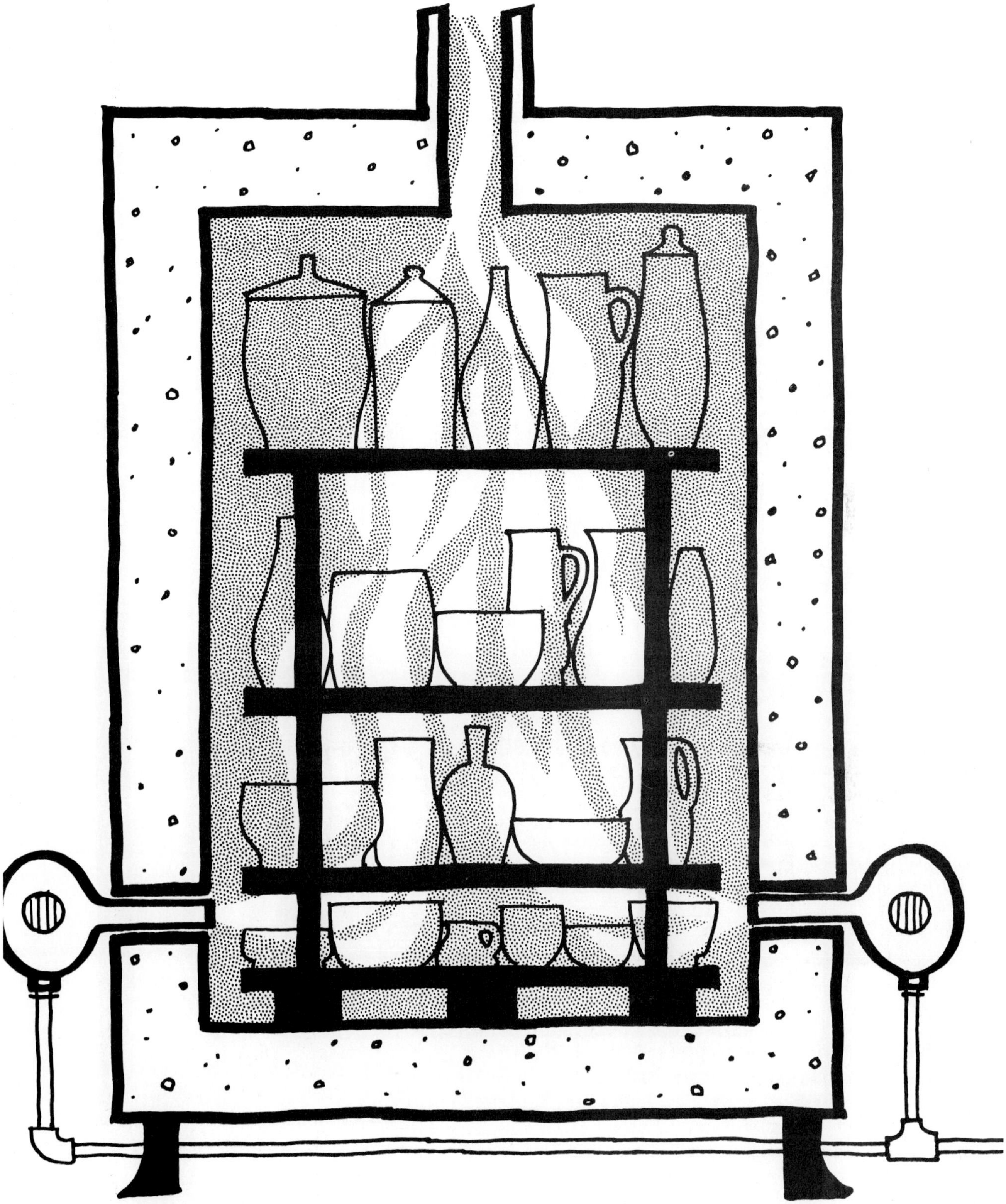

11-4 *An open updraft kiln.*

11-5 *An open downdraft kiln.*

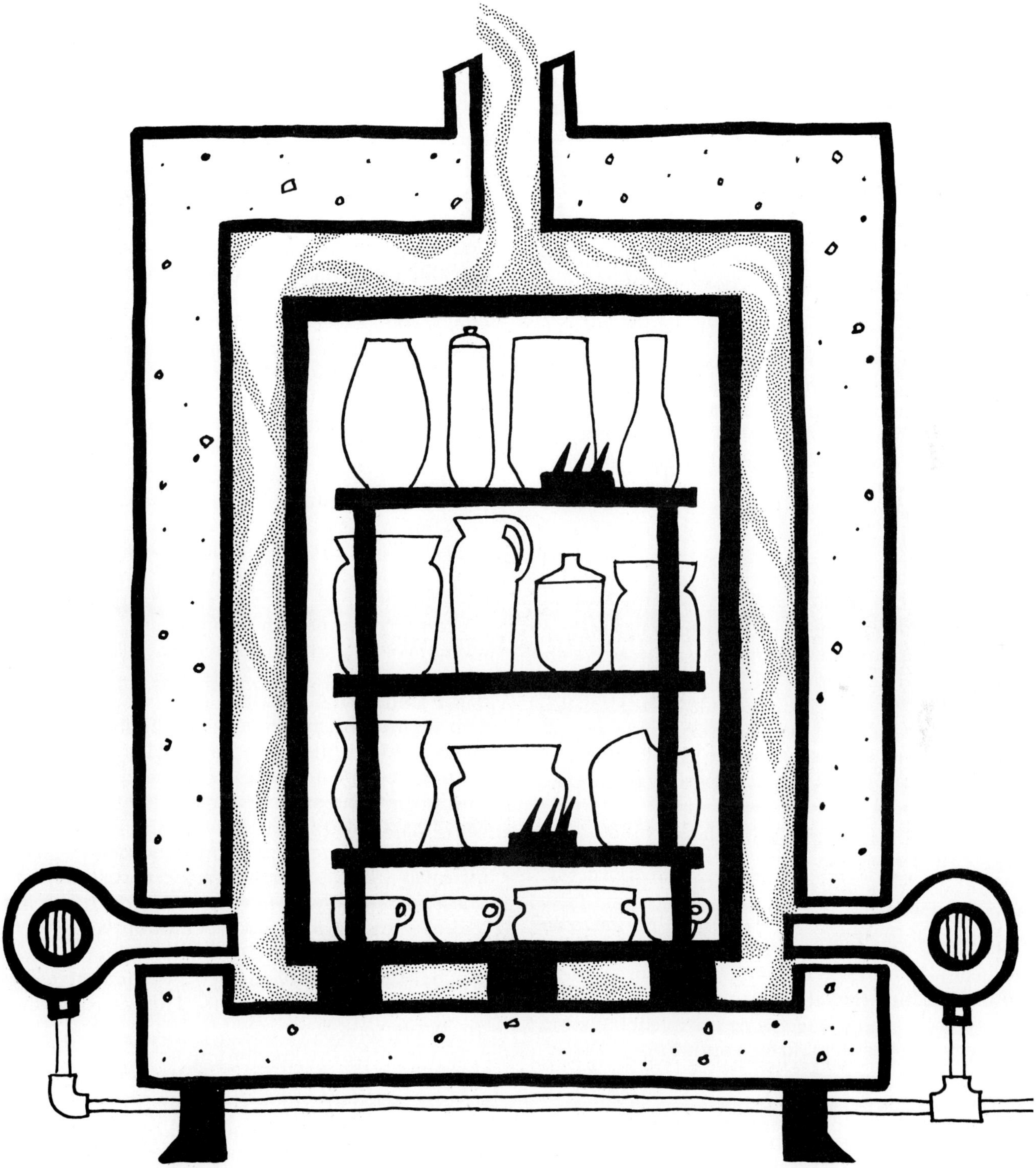

11-6 *A muffle kiln.*

potter stacks the pots accordingly, guided by a knowledge of the kiln's customary aberrations. A major advantage of the updraft kiln is its architectural concept, simpler to construct and less costly.

Open Downdraft Kiln Another variation on the open kiln derives from Chinese potters who, over a thousand years ago, realized the inadequacies of the updraft kiln. They changed the design and developed the downdraft kiln. The genius of the plan is that the flue is open at the bottom of the kiln wall. The consequence is to force the direction of heat flow from the vertical into recirculating patterns. Fire and hot gases churn around and pass over the wares many times before they reach the chimney opening at the kiln's bottom. Heat is more easily distributed evenly throughout the firing chamber, and fuel is consumed more efficiently while higher temperatures can be achieved. The kiln atmosphere as an active participant may be more effectively manipulated. Many potters who practice combustion firing prefer downdraft kilns, but they are more complicated in design and more costly to construct.

Muffle Kilns

The muffle kiln is designed to separate the active flame from the stacked glaze wares. This arrangement does for an entire kiln load what saggers do for a few pots. In the muffle kiln, clean atmosphere and uniform temperature are existing conditions. However, a drawback is the substantial waste of heat each time the kiln is fired, as the muffle walls or tubes receive heat, as well as the pots. Muffle kilns are infrequently seen in today's pottery studios.

HEAT SOURCES

Most firings take place somewhere within the range 1850° to 2400°. There are two practical methods to heat a kiln to the desired temperature. One is to burn a fuel, and the other is to fire with electricity. Even though there is no fire in electric kilns, the term applies to all firing conditions.

Combustion

Affluent cultures such as ours permit themselves the luxury of firing pottery with fuels that are uneconomical or inefficient, either for the pleasure derived from their use or for their perceived aesthetic and philosophic benefits to the work. Most other societies have burnt whatever fuels were available to them and were economically feasible. The Indians of the Southwest United States used dried animal wastes from the pastures. Potters in Ireland and Scotland burned peat. Coal was a favorite fuel in the late nineteenth and early twentieth centuries. Wood has been used where forest resources and cheap labor for cutting and splitting were at hand. In this country, studio potters today normally fire with gas and electricity. Some potters fire with wood or with other more exotic fuels—an anachronism in view of the technology available, but fully justified by the unique results.

For the studio potter, the major advantage of combustion firing derives from the nature of combustion. Air or oxygen combines with the fuel at ignition temperatures to give off heat and other products. The air inside the kiln is, therefore, an active agent in the firing process. Potters call this the kiln *atmosphere.* Chinese potters more than a thousand years ago first learned how to turn the combustion process and kiln atmosphere to their advantage, as an instrument of their art.

Oxidation firing is the outcome of an air supply at least adequate for a clean, hot burn. *Reduction* firing results when the air supply is diminished (reduced) so that, in a real sense, the combustion is starved for essential oxygen. The fire seeks oxygen to remain burning and finds it nearby in the clays and glazes of the pots stacked in the kiln. Oxygen is ever present in the pottery wares in the form of oxides, of which clays and glazes are constituted. Some of that oxygen is taken up by the fire and consumed in combustion. The consequences of this exchange are remarkable color and textural changes in the pots. The two most dramatic transformations take place when copper or iron are present, as colorants, in the glaze. Copper, when reduced, moves to its metallic state rather than to an oxide, towards the red side of the palette rather than the expected green. Having lost oxygen, iron is also converted into its metallic form from the original oxide, showing greys or grey-greens instead of the characteristic straws to rusty browns. These are probably the most notable reduction colors and are named accordingly: *celadon* for reduced iron and *copper red* or *Ox-Blood Red* for reduced copper. We generally consider the oxidation colors to be the more normal colors. They are described in Chapter 9.

In the firing process potters arrange for the atmosphere they wish to dominate in a firing. Everything else being equal, combustion firing seems peculiarly to give a heat somewhat more effective for melting glaze than the equal temperature in an electric kiln.

Electricity

Heat in an electric kiln is produced by the passage through a resistance of an electric current. The resistance most often used in studio and school kilns is a coiled, metal wire. Some electric kiln

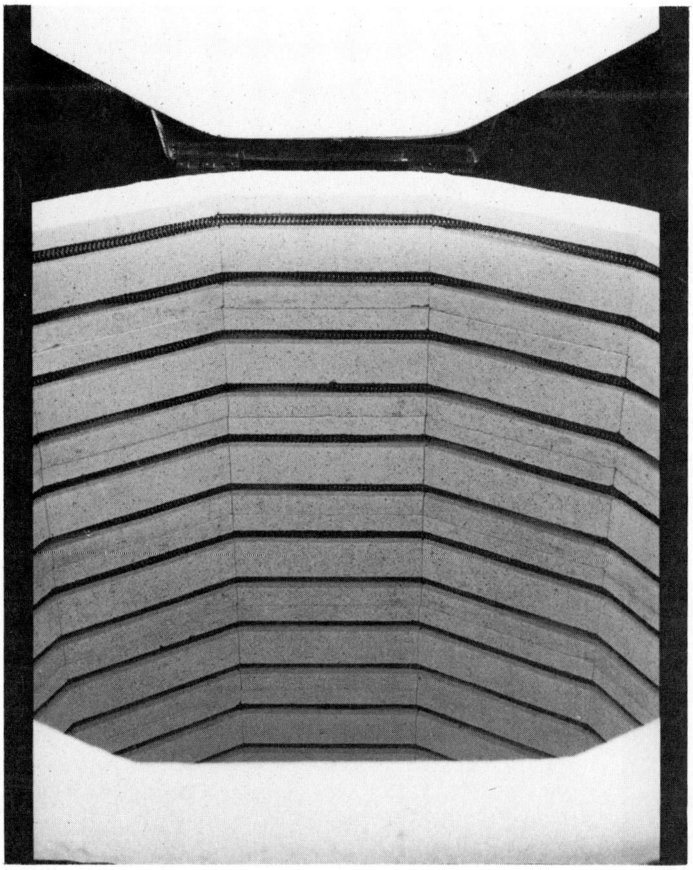

11-7 ***Coiled heating elements set in grooves in an electric kiln.***

Crusader Ceramic Equipment Corp., Indianapolis, IN.

models come equipped with silicon carbide rods as the resistances. The common term for either type of resistance is *heating element.* There are two types of wire elements:

Nichrome Wire Nichrome wire generally has a safe maximum firing temperature of about 1950°. The wire is wound in long coils which are embedded into channels cut into the kiln's walls and door. These coiled elements are controlled by switches mounted on the kiln to regulate the rate of temperature increase during firing. Given the practical maximum temperature of nichrome wire, only earthenware clays or glazes and low-fire bisking of stoneware and porcelain can be properly fired in one of these kilns.

Kanthal Wire Kanthal wire is made of an alloy which can safely be fired up to about 2400°. This type of wire is also wound into long coils, securely placed in grooves in the kiln's interior. Given the high temperature limit of kanthal, the insulating brick in the kiln must be at least as high in temperature value. Stoneware and porcelain might properly be fired in kilns with kanthal elements, as well as low-fire earthenwares.

Atmosphere The atmosphere in electric kilns is neutral or passive in its effect on the firing process. Theoretically, electric kilns could be heated in a vacuum; combustion kilns could not. Color and textural qualities in electric firing are quite similar to those of oxidation firing.

It is possible, in electric firing, to get close to the color qualities of reduction firing—not the same, but similar. The process is called *local reduction.* Several methods may accomplish this. One would be to add a small amount of 200-mesh or finer carborundum (silicon carbide) to the glaze while it is being weighed and mixed. Generally, one percent is adequate. During the firing, the carborundum substitutes as the agent to remove oxygen from the glaze oxides. With the proper conditions, glazes might resemble celadon, copper red, and other reduction transformations. Another method is performed during the actual firing, when the kiln is at least red hot. Through the peep hole introduce smoke-producing materials, such as moth balls or wet green twigs. Carbon will combine with oxygen from the glaze oxides. Again, it should yield local reduction effects. Potters who attempt these methods should have well-ventilated kiln rooms. Several trials may be necessary before a suitable combination of glaze, colorant, carborundum, application, and firing procedures bears fruit.

Kiln Types Electric kilns are widely used in the pottery community. Most high schools have only electric kilns, and many potters fire with electricity. They come in a variety of styles and sizes, with or without accessories for more convenient and

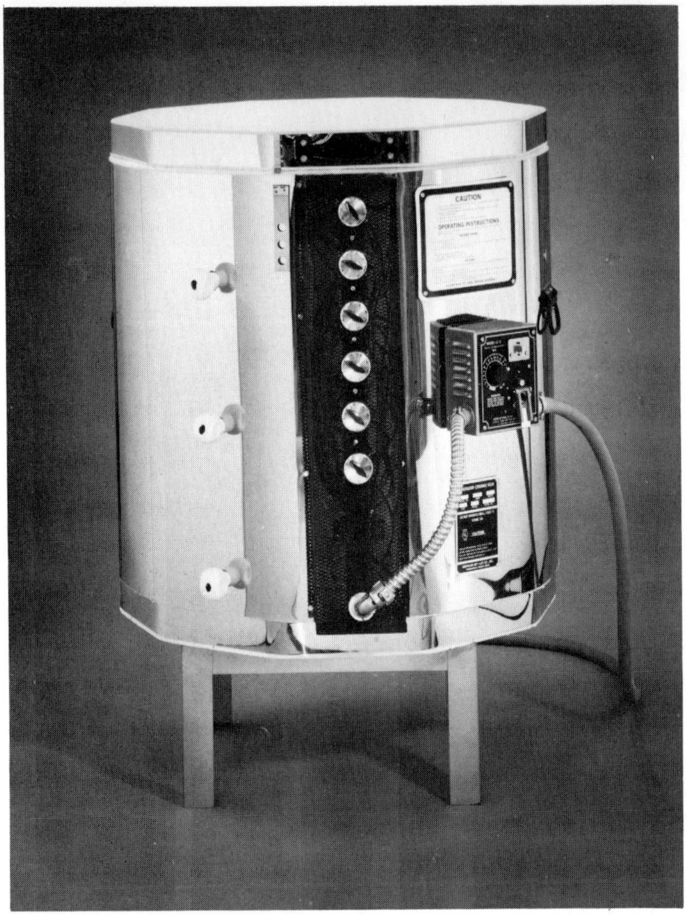

11-8 *A top-loading electric kiln, Crusader no. 274.*

Crusader Ceramic Equipment Corp., Indianapolis, IN.

11-9 *A front-loading electric kiln.*

A. D. Alpine, Inc., Torrance, CA.

precise firing control. The major difference found among electric kilns is where the door is located, at the top or at one side.

Top-loading kilns have doors which are hinged and open upwards from the top of the kiln. This type can be constructed more simply, without the structures needed to support a side-opening door. Therefore, top-loading kilns are bound to be less expensive. But there is a practical limit to the size of top-loading kilns because of the difficulties in handing pots down into deep firing chambers.

Front-loading kilns have doors hinged at one side which swing out sideways. Construction of such a kiln must be substantial to support the door's heavy weight. Therefore, these kilns are more costly. But they may be more easily stacked, even in large sizes. There is controversy over which of the two fires more effectively. Front-loading kilns are generally more generously endowed with wall and roof thickness for good insulation and heat retention. In some models of the top-loading kiln, there could be undesirable heat loss through the door at the top, which might be thinner. Potters are interested in a firing that does not take too long and in uniform heating throughout the kiln.

FIRING METHODS

There are only two ways to fire a kiln: from time to time or continuously. Studio kilns are of the first type: periodic, intermittent, or cyclical firing. From the very beginning, potters have fired their wares in this way. Of course, heat is lost each time the kiln is cooled and the door is opened, but the practical necessities in studio routine more than overcome the disadvantage. The kiln is stacked, fired, cooled, and emptied in cycles that suit the rhythm of work in the pottery studio.

In many pottery factories, large industrial kilns are fired continuously. Heat in the kiln is maintained at the desired temperature 24 hours a day, seven days a week. In these kilns, the wares are processed moving in a never-ending stream through the kiln, which is often in the shape of a tunnel. The wares enter at room temperature at one end, are brought to maturing temperature somewhere in the center of the kiln's length, and exit cooled at the opposite end.

12 HEAT MEASUREMENT IN THE KILN

Firing either bisk or glaze, it is essential to know when to turn off the kiln. The information should be readily available to the potter. It should validly relay what is actually happening to the pots inside the kiln. The most significant information about the firing is that the bisk is done and, in a glaze firing, that the glazes have melted well. When the potter is able to verify these conditions, the kiln is turned off and left to cool.

INDICATIONS FROM THE FIRE

Throughout history, potters have improvised many methods to affirm when the firing was properly done. Some of the methods are still practiced. Perhaps the most primitive is that the fire continues until tradition and experience convince the potter the work is done. This is not a particularly efficient method, for fuel conditions, the weather, atmospheric variables, and other circumstances are ignored by judgment based solely on the passage of time. Nevertheless, in the absence of more effective and sophisticated devices, this method of heat measurement is still used in some less developed pottery communities. Many beautiful pots continue to be made this way, as they have been for centuries.

Probably early on, potters observed that the color of the fire changed as the fire's heat grew more intense. They would have formed a direct association between heat and the color it produces. When the fire first shows incandescence at about 1000°, the color of the light is a dull, red glow, as in the early stages of sunrise. With rising temperatures, the color turns cherry red, orange, yellow, and perhaps almost white at about 2400°. With some experience and visual sensitivity, the potter learns to control the fire according to its color and knows that the wares are well done when the fire turns an appropriate color. Each time you fire a kiln, this valuable information will guide you in the increase of heat, even though some other device may actually signal you to adjust the heat or turn the kiln off.

Another method was developed from the observation that the higher the fire, the more a particular clay would shrink. If that same clay were fired once more under the same firing conditions, it would shrink an equal amount. A shrinkage test was developed using specially shaped pieces that had been left inside the kiln. During a firing they were periodically removed through a loose brick or trap door and measured against a calibrated scale. Finally, one such shrinkage trial would indicate that the desired shrinkage had occurred, and the fire had done its work so that the kiln could be shut down. This was a fairly sophisticated method for it relied on an objective and predictable measure, rather than subjective judgments. Some early potters used a method involving sample test pieces that were left in the kiln for the duration of the firing. Intermittently, one was removed and examined by the potter. Eventually, one examination would confirm that the bisk or glaze firing was satisfactory, and the firing ceased.

PYROMETRIC CONES

About 100 years ago a means for heat measurement was invented that remains in universal use today. The system is called *pyrometric cones* or *Seger cones*, after the inventor. We know them in the United States as 'Orton Cones,' for they were adapted for use in this country by Edward Orton. Originally shaped like small cones, they are now elongated, three-sided pyramids. They come in two sizes, the larger $2^1/_2''$ long, and the smaller just over 1″. Each is composed of oxides similar to the oxides that make up the clays and glazes. The composition differs from one cone to another, each cone being calculated to react differently to rising heat in the kiln.

Cones are placed standing up inside the kiln at a spot where they can be seen through the kiln's peephole. When the fire reaches a certain heat, the appropriate cone reacts to the heat by melting and bending over. The reaction of this cone is the sign the potter has been waiting for, and the kiln is turned off. In previous tests, each clay and glaze has been matched for its maturity with a cone equivalent, a cone that slumps just when that clay or glaze becomes mature. Potters always refer to cone values for clays and glazes, for example, cone 04 clay or cone 6 glaze and each clay or glaze in the studio has a cone equivalent. Purchased clays, bodies, and glazes come identified by their appropriate cone values.

For easy identification, each cone has been given a number. Cone 022 is the number of the cone that melts at the lowest temperature in the series, at about the point at which incandescence first appears in the kiln. The cone that takes a little while longer to melt at the next highest temperature is cone 021, followed in turn by cones 020, 019, 018 and so on until cone 01. The number of the cone that melts after 01 is 1. Cone 2 follows 1, then 3, 4, 5, 6, and so on until 42, the very last in the sequence. Each cone from 022 through 42 measures a step higher in temperature and an increase in the time needed to affect it. Cones, clays, and glazes all react similarly to the interaction of time and temperature, called by many potters the *heat-work*.

The cone numbering system is the result of historical development. The accompanying table should clarify the order and spacing of cone numbers. In actual practice, virtually all bisk and glaze firings take place somewhere within the range of cones 015–15. The others are rarely, if ever, used in pottery.

Most potters group three cones with the adjacent numbers (for example, cones 5, 6, and 7) pushed down into a single small coil, set at a slight angle (8°) and parallel to each other. This unit is called a *cone pack*. Set them only deep enough into the plastic clay to prevent them from falling over. The cones are lined up standing parallel in a row, almost touching. One flat side of each cone faces in the direction it leans to facilitate melting. The first cone is called the *guide cone* or *warning cone*. It is one number lower than the desired firing cone. The second cone, in the middle of the three, is the *firing cone*, the signal cone for the actual firing. The third cone is the *guard cone*, or *overfiring cone*. It is one number hotter than the firing cone. It should not melt if the firing is a good one.

For example, in a kiln whose wares require a cone 9 firing, use cones 8, 9, and 10. Cone 8 is the guide cone, the first to soften and bend over. A short time later cone 9 melts over completely and falls on cone 8, the signal to turn the kiln off. Cone 10 remains largely unmelted, but if by some mishap the kiln were fired too long, the degree to which cone 10 bends indicates by how much the wares are overfired. Many potters like to fire till the point of the guard cone softens somewhat, in the well-founded belief that just a little extra heat does the glazes some good.

Cones are efficient, effective, and inexpensive. Best of all, they give graphic evidence of the effects of heat on clays and glazes inside the firing kiln. We can tell by looking through the peephole from the cool safety outside, when the heat-work is done. The cone system is valid because cones are formulated from the same array of oxides as clays and glazes. They react similarly to the effects of time and temperature. If a kiln is fired at a faster rate, more temperature is needed to melt the cone; if fired less rapidly a lower temperature does the same work. Potters depend on the reliability of cones, for they operate identically time after time, and they do not malfunction. They are often used as a supplement and backup for other more complicated and expensive control instruments. Used cones should not be used again.

Pyrometric cones also react to atmospheric conditions during firing. Oxidation and electric firing normally melt the cones clearly and predictably, without unwarranted interference. However, in reduction conditions the cones may melt less well and show the effects of reduced oxygen on their surfaces. Therefore, a series of "iron free" cones is made available for reduction firing. All the standard cones are iron free cone 4 and above; they are white. Below cone 4 the color of the normal series is brown, due to the presence of iron in the formulation.

For use in small kilns, there is a series of small cones, sometimes called *junior cones*. They occupy less space in the kiln and in front of the peephole. Small cones are not precisely the equivalent of the standard size in that they require a few more degrees of temperature for them to melt.

TEMPERATURE EQUIVALENTS FOR ORTON STANDARD PYROMETRIC CONES*
AS DETERMINED AT THE NATIONAL BUREAU OF STANDARDS

CONE NUMBER	LARGE CONES				CONE NUMBER	SMALL CONES	
	60°C	108°F	150°C	270°F		300°C	540°F
022	576°C.	1069°F.	586°C.	1086°F.	022	630°C.	1165°F.
021	602	1116	614	1137	021	643	1189
020	625	1157	635	1175	020	666	1231
019	668	1234	683	1261	019	723	1333
018	696	1285	717	1323	018	752	1386
017	727	1341	747	1377	017	784	1443
016	764	1407	792	1458	016	825	1517
015	790	1454	804	1479	015	843	1549
014	834	1533	838	1540	014	870	1596
013	869	1596	852	1566	013	880	1615
012	866	1591	884	1623	012	900	1650
011	886	1627	894	1641	011	915	1680
010	887	1629	894	1641	010	919	1686
09	915	1679	923	1693	09	955	1751
08	945	1733	955	1751	08	983	1801
07	973	1783	984	1803	07	1008	1846
06	991	1816	999	1830	06	1023	1873
05	1031	1888	1046	1915	05	1062	1944
04	1050	1922	1060	1940	04	1098	2008
03	1086	1987	1101	2014	03	1131	2068
02	1101	2014	1120	2048	02	1148	2098
01	1117	2043	1137	2079	01	1178	2152
1	1136	2077	1154	2109	1	1179	2154
2	1142	2088	1162	2124	2	1179	2154
3	1152	2106	1168	2134	3	1196	2185
4	1168	2134	1186	2167	4	1209	2208
5	1177	2151	1196	2185	5	1221	2230
6	1201	2194	1222	2232	6	1255	2291
7	1215	2219	1240	2264	7	1264	2307
8	1236	2257	1263	2305	8	1300	2372
9	1260	2300	1280	2336	9	1317	2403
10	1285	2345	1305	2381	10	1330	2426
11	1294	2361	1315	2399	11	1336	2437
12	1306	2383	1326	2419	12	1355	2471

*When heated at the rates indicated.

Courtesy of Edward Orton Jr. Ceramic Foundation.

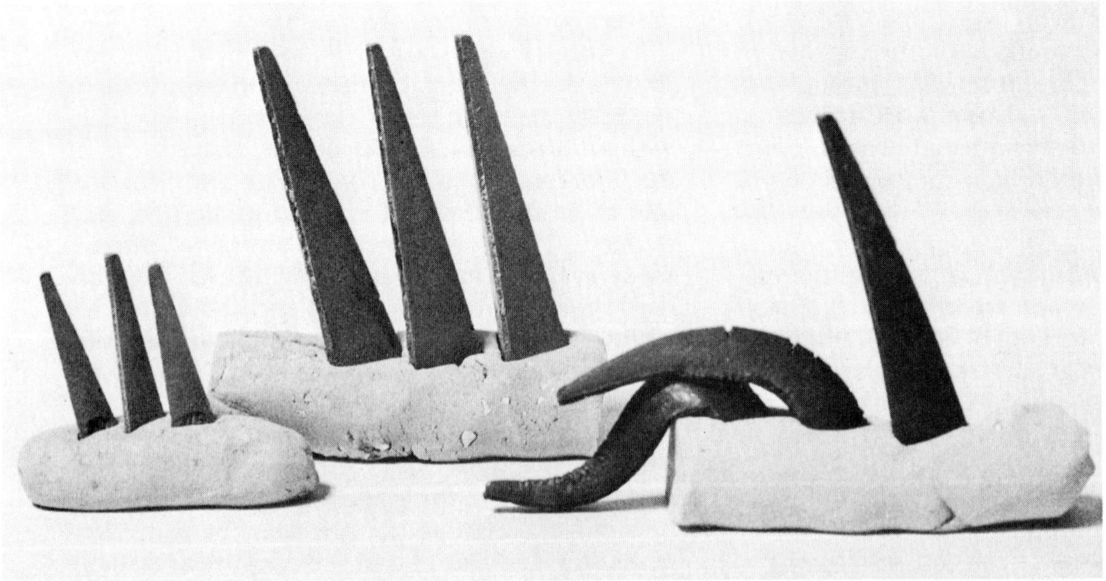

12-1 *Pyrometric cones. From left to right: small cones, unfired; large cones, unfired; large cones, fired.*

ELECTRIC PYROMETERS

An electric pyrometer is an instrument which indicates at a glance the actual temperature in degrees inside the kiln at any given moment. Pyrometers, unlike cones, do not reflect the effect of time and temperature on the work, merely the temperature. The indicating dial is generally mounted directly on the kiln, although there are portable electric pyrometers available. Two wires lead from the dial through the kiln wall, connected to a probe inside the kiln. The probe is called a *thermocouple.* When the kiln heats, the thermocouple sends an electric message, indicated on the dial face in terms of degrees of temperature. Electric pyrometers are not plugged into a wall socket for electricity; rather, their electricity is self generated when the thermocouple is heated.

An *indicating pyrometer* merely shows the temperature inside the kiln. It is an extremely useful instrument during the course of firing, for it informs the potter how fast the temperature is increasing. Armed with this information, the potter can make appropriate adjustments to regulate the advance of heat more effectively. Indicating pyrometers are not expensive. Most kilns come with them, or they are a recommended accessory.

Controlling or *shut-off pyrometers* are electric pyrometers that have the capacity to turn off electric or gas kilns automatically at predetermined temperatures. The potter sets the instrument for a desired temperature; the firing stops when that point is reached. It is a convenience, but there are some disadvantages: (1) pyrometers sense temperature only and not the entire time-temperature interaction, the effective combination for all firing, and (2) pyrometers are mechanical-electrical devices subject to failure, too often at the crucial moment. Some potters use this instrument solely as a safety back up to prevent overfiring.

AUTOMATIC SHUTOFFS

Available as accessory instrumentation on electric kilns, these are devices that turn the kiln off when the appropriate cone has melted. The shut-off devices combine a pyrometric cone with an automatic switch. The potter places a cone on one end of the instrument inside the kiln. When the cone melts, it activates the instrument to stop the kiln from firing further. A number of different brands are available. Seemingly a valid contribution to convenience in firing kilns, it may nevertheless also malfunction at the crucial time. The reader would be well advised to use a full set of cones along with any of these instruments, as a reliable indication of when the work is properly done, and it is time to turn the kiln off. Sophisticated electronic automatic shutoffs are also available for gas kilns. Potters should exercise the same caution given above when these instruments are used.

13 STACKING AND FIRING KILNS

STACKING A BISK KILN

A bisk kiln is properly stacked when it has been filled with pots arranged in a pattern of uniform density, and each piece is set in place without risk to itself or to its neighbors. The stacking must be performed carefully and methodically.

The first step is to prepare the cones for firing (see Chapter 12). The prevailing practice among potters is to bisk fire at a cone low enough for the fired clay to remain porous and absorbent, yet high enough for the work to be hard for safe handling during the glazing. Absorbent bisk is easier to glaze because liquid glaze adheres more readily to porous than to impervious clay. Much bisk firing for medium- and high-fire wares, such as stoneware and porcelain, is to cone 07 or 06. Bisk firing for low-fire earthenware might be better at cone 08 or 07.

The greenware should be completely bone dry (see Chapter 10). Study the shape and size of the kiln and make an estimate as to which pieces, by virtue of their sizes, forms, or functions, will be selected for this firing. In small kilns, the stacking arrangement will be more secure if the shorter pots are placed at the lower levels in the kiln and the taller pieces towards the top. Select pieces of fairly uniform height for each shelf.

For convenience and safety, the practice is to place the pots on shelves, although in earlier times wares were piled together in a heap in the kiln. While arranging pots on the bottom shelf, leave space for three posts of equal height, but about $^1/_2$″ to 1″ above the tallest pot on that tier. When the first level is filled, place the second shelf carefully on the three posts, and continue to stack that shelf. Keep on in this manner until the kiln is completely stacked.

Shelves and posts come under the heading of *kiln furniture,* the paraphernalia of the kiln for stacking pots. Black shelves are made of silicon carbide, very refractory and suitable for firing to high stoneware and porcelain temperatures. Buff or tan shelves are generally manufactured of cordierite, which is less refractory and should not often be used above cone 6.

Large shelves are extremely heavy and difficult to lift. Therefore, when stacking big kilns with a sizable floor area, potters often resort to using half shelves or quarter shelves, as well as odd sizes. It is best, under these circumstances in a front-loading kiln, to first stack the entire back from bottom to top, then the front from bottom to top.

Stack bone-dry greenware as close together as possible, touching where feasible. However, separate the pieces from kiln walls and kiln furniture, about $^1/_2$″ inch to 1″ distant. Pots are safer standing on their own feet rather than being laid on their sides. Where shapes permit, nestle pots inside each other, lighter and smaller pots inside larger and heavier ones. The pieces inside should rest their feet on the interior base of the supporting pot, clearing the walls. When series of pots have been made, such as mugs or bowls, they may be stacked totem-pole fashion, rim to rim and foot to foot. Surprisingly, there is no unbearable strain on the bottom pot. Pieces with lids are arranged in the

13-1 *A kiln stacked for bisk firing.*

13-2 *Pots arranged for bisk firing.*

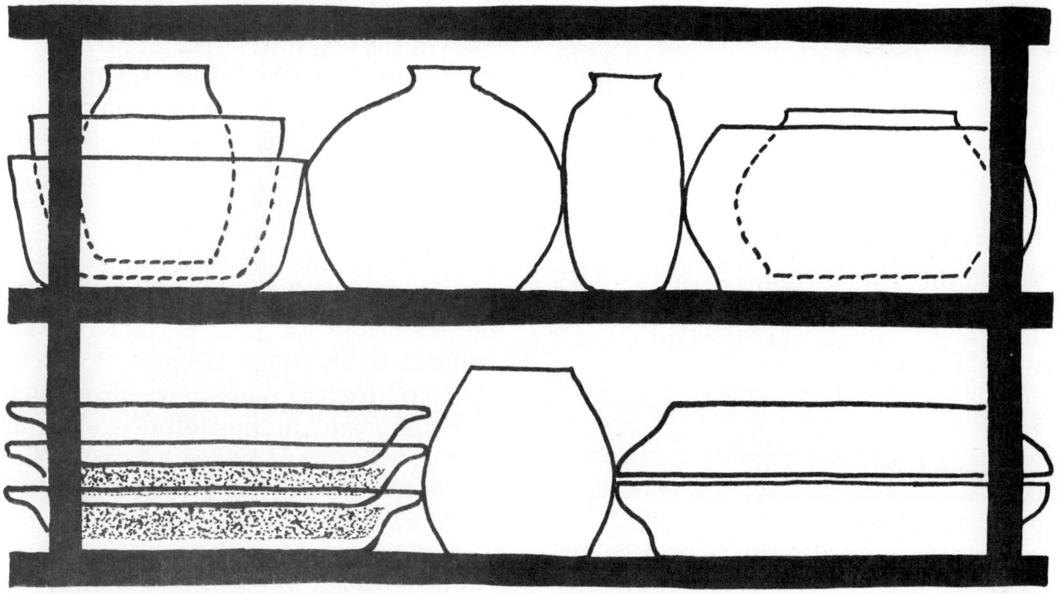

13-3 ***Cones viewed through the peephole in the kiln door.***

kiln with the lids in place on their pots. Lids with tall or fragile handles may be inverted in the pot for safety's sake and to conserve space.

Some pieces require special handling, because they are large and heavy or because of their shapes. Massive pots and sculpture stand a better chance of coming out well if the potter first sprinkles a light dusting of powdered flint or fine grog on the kiln shelf where the pot is to rest. The gritty material acts as a ball-bearing surface underneath the heavy weight, permitting it to move while shrinking during firing. Strain that could otherwise cause cracking is minimized. Sometimes, it seems best to fire some forms, such as large bowls, on their rims, and a similar dusting of flint or grog is imperative.

Flat shapes, such as plates and platters, may be stacked nestled inside each other. However, there should be an ample bed of fine grog or clean sand laid in each to support the next piece.

In a good stacking, the potter has filled the kiln space to the walls and ceiling, leaving only about 1″ of clearance. The pieces have been arranged to avoid empty gaps or air spaces. Heating will, as a consequence, be more uniform, important in bisk firing and critical in glaze firing.

The last step is to place cones on a shelf, where they can be seen through the kiln's peephole. Two sets of cones may be used in large kilns, where there are two peepholes, one near the bottom and one at the top.

FIRING A BISK KILN

Bisk firing transforms the clay into a permanent, hard, rock-like material. There has been an irreversible chemical change, the clay can no longer be made plastic when mixed with water. Firing shrinkage occurs, and the clay assumes its true color. Bisk firing is an essential process to make pottery serviceable.

Water Smoking

Once the kiln has been stacked and the cones set in place, firing commences. The first setting in an electric kiln is low; it is a very slow start in a combustion kiln. Temperature increases at a leisurely rate, about 150° to 200° per hour. An indicating pyrometer is a useful instrument to track heat rise, until the cones take over towards the end of the firing. During the first several hundred degrees, until about 300°, the major effect is to drive all residual moisture from the clay, some of which has been retained within the clay structure even though it is bone dry. Pore water and atmospheric humidity, deep within the clay's composition, are changed into steam by temperatures above 212°. The water vapor makes its way into the kiln's atmosphere and then out through openings in the kiln. Therefore, this first firing stage is aptly named *water smoking.* Were water smoking done too rapidly, the pot might react violently to the force of steam and break or even explode. It is an unhappy moment when the kiln door opened reveals "popcorn" and "shrapnel" where once a proud pot had stood, but it is good that this is about the only occasion when pots can blow up.

Dehydration

After water smoking, the heat input may be increased somewhat. During this next firing stage the clay molecule is transformed by temperature high enough to act as catalyst, from its unfired to its fired composition. The chemical change is irreversible. By now the temperature has climbed to above 700°, high enough for the organic ingredients in the clay mass to begin to burn. These combustible materials had been combined by nature with the clay at the time of deposit in the earth. The temperature continues to increase and now has reached above 1000°.

At 1063° a natural phenomenon occurs, unavoidable and of considerable consequence to the potter. It is called the *quartz inversion.* All clays contain a portion of silica or quartz as a residue of their origin as rocks. At 1063° the silica expands somewhat, about one percent. Upon cooling of the kiln, contraction back to the original size occurs. If a significant proportion of additional or accessory quartz should be present in the clay or if cooling has been too speedy, the strain of expansion and contraction may be too much for the pot. There is the danger of cracking. This second stage of firing is called *dehydration,* named for the irreversible chemical transformation. It continues until about 1400°.

Oxidation

Oxidation is the third firing stage. During this period, which lasts for about 200°, all the remaining combustible, organic impurities in the clay are burnt out. The gases of combustion leave the clay and enter the kiln's atmosphere. Electric kilns in enclosed, unventilated spaces may emit smoke and acrid fumes at this time, all the more reason for well-ventilated kiln rooms and studios. During this period, firing should not be too rapid, for these gases must be permitted safely to percolate from the clay's pores. By the time all of this has taken place, the temperature in the kiln has risen to about 1600°.

Vitrification

Vitrification is the name of the final stage in bisk firing. The clay undergoes firing shrinkage and develops its true fired color. The kiln is turned off when the firing cone bends all the way over, its tip touching the guard cone. A few minutes earlier, the guard cone had given its warning by melting. Exercise care looking through the peephole. Sunglasses may assist in seeing the cones through the bright light of the fire. An over-eager potter might get too close to the peephole, and eyelashes are singed. Sometimes the light inside the kiln is hazy from the fumes, and the cones are difficult to make out. Gently puff a breath of air in through the peephole, and look quickly.

Leave the kiln door closed after turning it off. Now the potter patiently waits for the kiln to return to room temperature and become cool enough for the wares to be comfortably handled. If a kiln has been fired for about eight hours, it may take overnight or longer to reach safe cooling. The kiln door may then be opened and the pots carefully removed and stored to await glazing.

STACKING AND FIRING A GLAZE KILN

Glazed pots are particularly vulnerable to conditions stemming from the arrangement and stacking in the kiln. Variations in heat and atmosphere leave telling marks on the glaze in terms of color, texture, and opacity. The pieces should be carefully located in the kiln for uniform distribution; no large, empty spaces in one section while the pieces are crowded together elsewhere. Generally, the method of stacking is much the same as for bisk firing, but there are some special preparations.

Kiln Wash and Shelves

First, prepare a cone pack, and set it aside to dry. The floor shelf of the kiln and all other shelves must be coated on their top surfaces with *kiln wash.* This is a protective covering applied to the shelf in the event that some glaze might, during melting, run or drip down off the pot onto the shelf, with risk to both. If the shelf has kiln wash on it, glaze spills can then easily be removed by scraping or chiseling. The potter keeps a chisel and mallet on hand in the kiln room for this purpose.

There are a number of kiln-wash formulations, but a good, general purpose mixture is 50% kaolin and 50% flint, combined with enough water for a paint-like consistency. Kiln wash may also be purchased ready made from pottery-supply firms. Brush it on the shelf surface, being careful not to drip over the edges or coat the underside. Brush an even coating, so that the pots have an unbroken surface to sit on. Unless the shelves are brand new, they

13-4 *Glaze drips on kiln-washed shelves can be easily removed.*

probably already have kiln wash on them from previous use in glaze stacking. Kiln-wash coatings that are clean and level may be used over repeatedly. However, glaze drips and buckled, crusty surfaces should be removed by scraping and chiseling. Bare spots can be touched up as needed or an entirely new layer painted on.

In most studios the same kiln shelves serve both bisk and glaze stacking. Greenware can safely and confidently be placed on washed shelves if the coating is clean of glaze drips and the wash surface is level. However, for all practical purposes there is no need for kiln wash when stacking for bisk firing.

Stacking Glazed Wares

Select pots of uniform height for the same shelf. In small kilns the arrangement will be more stable if the shorter pots are on the lower levels, leaving the taller pieces for the top tiers. Footrings must be completely bare of glaze. Space the pieces about 1/2″ to 1″ apart, to avoid having them fuse together when the glazes melt. For the same reason, lids would be stacked on shelves next to their pots. Lids can be stacked on their pots only if the tangential, contact surfaces are completely clean of glaze.

Dried glaze coatings are extremely fragile. Careless handling may chip glaze from rims, handles, and other projections, and some of it could be rubbed from the sides of the pots. These are difficult, if not impossible damages to repair, and the potter is often forced to wash the old glaze from the piece and apply a new coating. If glazed work has been sitting around for some time, it is a good idea to blow away any studio dust that might have accumulated on it.

There is always the concern when firing glaze that some might run down over the pot's foot to

13-5 *Glazed pots should be stacked apart from each other.*

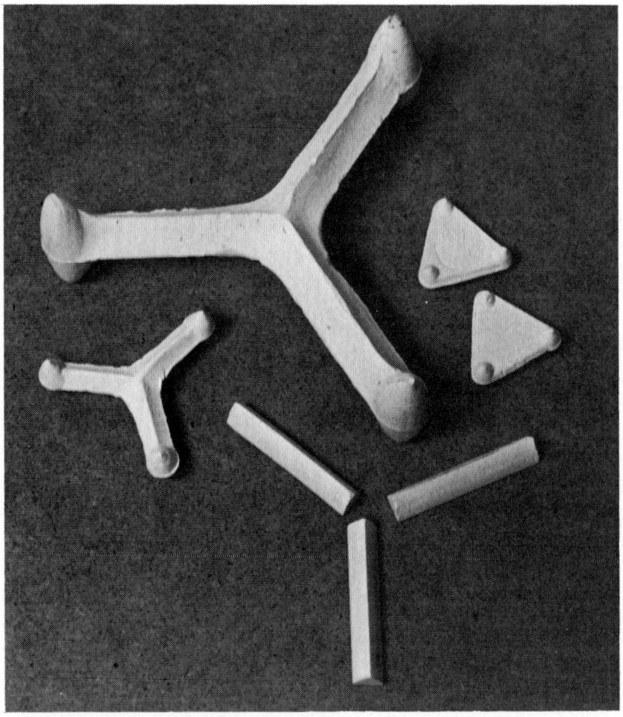

13-6 *Stilts, spurs, and pins.*

the kiln shelf so that upon cooling the pot is found to have fused to the shelf. Kiln wash helps protect shelves from damage. Clean footrings add another measure of safety to the pot and the shelf, as does careful glazing. However, there is still risk of running glazes from coatings too thickly applied, overfirings, and miscalculations of glaze recipes.

For additional safety, many potters use stilts, spurs, or pins placed under the footrings of glazed pots. They support the piece and raise it away from the shelf. Should glaze unfortunately run, the shelf at least will be protected. Kiln furniture is expensive. Spurs are the smallest and are usually used for tiny pots and jewelry. Stilts come in a wide range of sizes. They give the most protection when their points rest directly under clean footrings. For very large and heavy forms, as many pins as required are arranged under the feet or bases. These useful little objects are manufactured from a special clay body. They are also made with wire points for low-fire wares.

Glaze Firing

Glaze firing may be done more rapidly than bisk firing. However, as the firing is usually to a higher cone, it is not a significantly shorter procedure. The guard cone melts and bends to give warning. In a few minutes the firing cone also melts and keels over. Many glazes benefit if the firing is extended a short while longer, until the tip of the last cone is seen to soften and slightly bend.

Many potters prefer to continue the heat at the end, for what is called *soaking*. The objective is to keep the kiln steady at the high terminal temperature for a brief period, perhaps 15–30 minutes. Soaking may be at a slightly lower temperature, but it should not result in the last cone melting further. Potters hope that by soaking, the glazes will more perfectly melt, and some flaws will be minimized or eliminated—this usually occurs. Soaking is not difficult to control in an electric kiln. Turn the switches down to "medium" or "low" rather than "off." Watch the cones and pyrometer carefully, to maintain an even temperature not higher or lower than appropriate.

After the firing is completed, the door is left closed. Permit the kiln to cool slowly. Too rapid cooling may be the cause of glaze cracks and pinholes. Glazed pieces may be taken from the kiln only after they have become cool to the touch. At first open the kiln door partially, to allow the warm contents

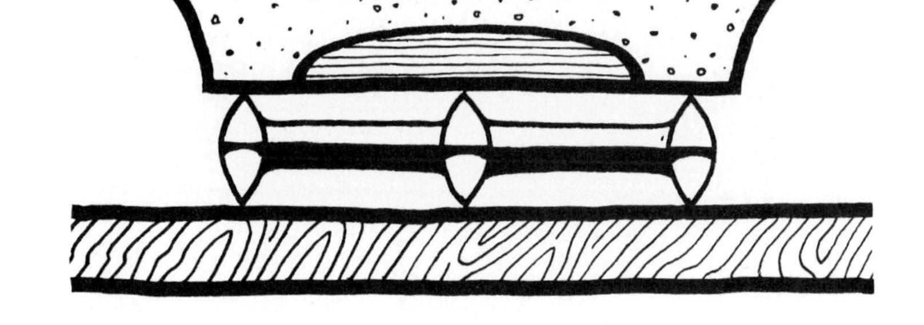

13-7 *Support the pot directly under a clean footring.*

some time to cool further. The potter's patience is tried waiting to get into the kiln. But at the end of the waiting period, the reward is the recurring and incomparable experience when the finished work is brought to light.

THREE FIRING PRINCIPLES

There are three conditions in all firing that guide the potter to successful work: firing point, firing rate, and atmosphere.

Firing Point

Each clay, body, and glaze has a cone equivalent to indicate the most suitable heat. At their best, clays and bodies have been fired as high as possible, stopping short of their melting, slumping, and bending. Desirable qualities of hardness, imperviousness, durability, glaze fit, color, and texture are fully developed when the wares are fired to maturity. Glazes also melt more perfectly at appropriate cones. Some glazes melt well at only one cone, but the more useful glazes have a wider temperature range. Clays and glazes purchased prepared are always identified as to how high they should be fired. Mixtures whose cone values are unknown can be subjected to easy tests to determine their proper firing ranges. Fire the sample first at a very low cone, then gradually increase the cone number in subsequent firings, until the sample responds appropriately. (Tests for clays and bodies are given in Chapter 4.)

Firing Rate

The mechanical and chemical interactions of pottery materials do not permit the pot to survive swift firing. Most successful bisk and glaze firings in smaller electric kilns take in the neighborhood of six to eight hours. Firing larger kilns and gas or wood-burning kilns can easily last much longer. It is important in bisk firing for the water smoking and oxidation stages to proceed slowly so that the gases generated leave the clay slowly. In glaze firing the final period should be slow for the best results.

Atmosphere

The atmosphere in electric kilns is neutral or passive as a factor in firing. However, at times an excessive accumulation of gases given off by the clay and glaze chemicals affects color. It is an undesirable reduction atmosphere, which, fortunately rarely occurs. Normally, the potter may confidently ignore atmosphere in electric firing. At the same time, firing with electricity removes atmosphere from acting as an agent to create unique color qualities. When firing combustion kilns, atmosphere cannot be ignored, for it is an essential and active component of the process and one of the tools of the trade. The potter learns how to control and bend atmospheric condition to fruitful ends.

14

GLAZES AND GLAZING

WHAT IS A GLAZE?

A glaze is a silicate mixture fused by heat into a glassy coating on clay surfaces. Glaze is actually a type of glass, defined technically as a *super cooled liquid.* It is applied on pots as a mixture of oxides with water, a creamy liquid which has no resemblance to its final melted and cooled appearance. Glaze formulations are like those of glasses in that silica is the essential and major ingredient. Other oxides in the glaze mixture make diverse contributions, such as determining how the glaze will melt and react during cooling and what its ultimate visual and textural properties will be.

Technical and aesthetic qualities of glaze can be arranged to suit the potter's requirements. This is accomplished by selecting for the compound those oxides that are needed and determining their proportions. How high the glaze is fired, and other circumstances surrounding the firing process, are other means to develop variety in glaze quality. Another method to achieve variation concerns how the glaze is applied. In the actual practice of creating a glaze, the oxide composition, firing temperature, and kiln conditions are adjusted to each other for technical perfection and aesthetic vitality.

Most studio glazes melt well at an appropriate temperature within the range 1800° to 2400°. Some glazes are formulated to mature over a broad spread of 100° to 200°, but most glazes seem to do best when fired within a narrower range of some 20° to 50°.

GLAZE PURPOSES AND GLAZE DIMENSIONS

Glazes function in three basic ways:

1. **They make the pot waterproof.** They prevent penetration of liquid into the fired clay ware. Impervious glaze permits pottery to be efficiently and safely used for the storage, preparation, and consumption of foods and other necessities. This may actually be the most useful function glaze has contributed to mankind.

2. **They present a smooth, long wearing, and easily cleaned surface.** Glazes furnish a sanitary, disease-free coating, fused permanently to the clay.

3. **Glazes provide infinite opportunity to enrich the aesthetic properties of the pottery form.** In what ways do we sense a glaze? Which glaze dimensions reach our consciousness? There are three fundamental visual qualities in any glaze:

a. **Color** is probably the most evident tool potters have to influence a glaze's appearance. Color is a rich source of visual material for the pot, made even more productive when the potter comprehends the uses of color's components: hue, value, and intensity. (See Chapter 9 for a thorough description of color in glaze.)

b. **Texture** is the tactile response to glaze surface. All pots are at least visual objects, and most of them are meant to be held and handled. Every glaze has one of a variety of possible textures captured in its surface. On some pots there will be glaze textures arranged in abundance for contrast and variety in the composition. Other forms are more appropriately glazed in a single texture. Some glazes are very shiny. The potter's traditional name for such a highly reflective surface is *bright.* At the opposite end of the texture scale are less reflective qualities, dull, *matt glazes.* Formulas can be adjusted to produce matt surfaces. Bright glazes can be altered for dull results by the addition of small amounts of selected oxides. Some matt glazes are extremely dull, even roughened, pitted, or streaked. Such a matt glaze might be very attractive to look at or touch, but it would not be suitable for a pot which is intended to hold food, to be washed, or to be held to the lips.

Other matt glazes are silky, satiny, or unctious. Probably the more traditional, most often used description of a matt glaze is that it has a soft and buttery surface. It makes an unmistakable, strong appeal to the touch. There are glazes that fall somewhere within the opposites of bright *versus* matt. These are called *semi-matt glazes.* Many potters are stimulated to compose interesting surfaces more by texture than color. Texture is the more subtle of the two and, therefore, may very well be the more challenging and rewarding design instrument.

c. Transmission of light is the glaze dimension which, of the three, is the most directly related to glaze's origin as glass. When glass or glaze is perfectly melted, the result is a coating through which there is no interference with the transmission of light. We can see clearly through the transparency to whatever there is visible in the clay's surface. Potters call transparent glazes *clear glazes. Opaque glazes* are those through which no light passes, which completely obscure that which is underneath. Opacity results when selected oxides, deliberately introduced for the desired effect, do not completely melt within the glaze mixture. They remain "floating," interfering with the passage of light. These special oxides are called *opacifiers.* Between the extremes of clear and opaque, there is a potentially wide range of semi-opaque and semi-transparent glazes. Some pots are made lovely by the interplay of translucencies, cloudy, milky, and veiling here, revealing there.

The study of glaze should be approached in terms of the three basic components: degree of opacity, textural variation, and color quality. The potter can then make more sensitive and satisfying discriminations when glazes are chosen for the work. The color effect of transparent glazes to which color has been added is considerably modified by the color of the clay beneath the glaze. For example, blue in a transparent glaze does not appear blue when applied over red clay. Glaze color shows up most nearly its true self in an opaque glaze or over white clay. Matt glazes are, as a by-product of the chemistry which causes mattness, somewhat opaque; there is no such thing as a completely clear matt glaze. Conversely, clear glazes are bright. Transparent glazes made opaque by the addition to them of opacifiers remain bright.

Potters may appear to be preoccupied with the visual aspects of glazes. However, for the work to be entirely successful, the potter must have sufficient command over glaze technology: glaze composition, preparation, application methods, and behavior of the glaze in firing. How will a glaze function in everyday use, in reaction to temperature change, food acids, abrasion and wear, and similar physical stresses?

GLAZE COMPOSITION AND FORMULATION

Glazes are prepared by weighing, mixing, and grinding silica with other inorganic materials. How a glaze behaves in application, melting, and use depends primarily on which of these oxides are present and what their proportions are in the compound. Fewer than one dozen oxides comprise the total list from which ingredients for most glazes are chosen. Many glazes consist of only six or less. Nevertheless, the possible permutations are practically limitless.

Glazes may be formulated from scratch by the potter or weighed according to established recipes that are begged, borrowed, or stolen. Probably the easiest way to have a good glaze is to purchase one, already prepared and ready to apply to your work. No matter how obtained, all glazes can be broken down to their ultimate oxide fundamentals.

Glaze Calculations by the Empirical Method

A good way to create glazes is by the empirical formula method. This procedure is much utilized by full-time potters and as such deserves at least an introduction in this book. Beginning potters might be better off acquiring proven glaze recipes or using prepared glaze mixtures, budgeting time and energy for more central achievements in art and craftsmanship. For the reader's information and for future reference, a brief description of the empirical method follows.

The oxides which most often appear in glazes are selected from among the following:

PbO	lead oxide
CaO	calcium oxide
ZnO	zinc oxide
BaO	barium oxide
Na2O	sodium oxide
K2O	potassium oxide

MgO	magnesium oxide
Al203	aluminum oxide
B203	boric oxide
SiO2	silicon dioxide

Any of these oxides may be selected for its contribution to the glaze mixture. Silica and alumina are basic and present in almost every glaze.

Numerical quantities are assigned to each oxide chosen for the glaze. The number represents relative quantities (molecules) in the glaze, signifying a proportion of the total. The oxides are then arranged in three groups or columns, according to their element–to–oxygen ration. The three groups are:

RO	R2O3	RO2
1 radical plus 1 oxygen	2 radicals plus 3 oxygens	1 radical plus 2 oxygens

(Note: **Radical** means the element.)

The resulting glaze statement is an empirical formula, which may be written as in this example, a low fire clear glaze:

RO	R2O3	RO2
PbO .7	Al203	SiO2
CaO .2	.25	1.25
ZnO .1		
1.0		

The empirical formula is the factual description of the fired glaze. Were the fired glaze analyzed, it could be described in the manner shown above. It is a statement of the essential, nutrient oxides in the glaze.

It is impractical, inconvenient, and expensive to mix the glaze directly from the arrangement of oxides in the empirical formula. Therefore, the formula is converted into a useful recipe, which is a list of the practical ingredients needed to furnish the desired oxides. Each recipe ingredient furnishes one or more of the oxides in the empirical formula. In this example white lead supplies PbO; whiting furnishes CaO; feldspar supplies K20, Na20, Al203, and Si02; kaolin is chosen for Al203 and SiO2, and flint for Si02. The quantities are calculated arithmetically and the resulting glaze recipe written in a form similar to the following:

White lead	130
Whiting	20
Feldspar	56
Kaolin	29
Flint	33
	268

These are the ingredients and the quantities actually weighed in grams and mixed by the potter and the method used by pottery-supply companies to develop the glazes they sell. Glaze ingredients are readily available in the form of dry powders, purchased by the pound from pottery-material suppliers. Most of them are inexpensive.

Recipes should be tested for workability prior to application on good pots. Representative samples are weighed, mixed, and fired on test pieces. Results might indicate adjustments of one kind or another, which are first made in the empirical formula. The recipe is changed accordingly, and the new version is tested again until a satisfactory glaze results. The whole procedure can be fascinating work, in the spirit of experimentation and invention, but the beginning potter is urged to postpone this category of activity for a more appropriate time.

GLAZE TYPES

All the qualities of a glaze develop out of complex interactions involving the clay body, the type and amount of oxides present in the glaze, how the glaze is applied, variations in heat, and how the kiln is fired. Consequently, an infinite variety of glaze effects are possible. It is easier to comprehend the whole business of glazes if we place them in categories according to their common traits. In this way we can come to grips with quantity and variety, more effectively study the influences affecting them, and more sensitively choose the right glazes for our pots. Potters use different systems to describe glazes; there is no single accepted classification method. The following method is at least as helpful as any other. It is based on five criteria: *how the glaze is prepared for use, its composition, cone values, texture, and ability to transmit light.*

Glaze Preparation

Raw glazes. These are glazes prepared substantially from raw materials. A raw material is a recipe ingredient that, except for being pulverized, is much the same when mixed in the glaze as it was in the earth. Most raw materials are earthy minerals, although some metallic oxides appear in glaze recipes. Some of these ingredients are cleaned and made uniform but not otherwise changed or processed. Nevertheless, from year to year many raw materials vary somewhat in their chemical composition, due mostly to differences in the natural deposits. Glazes may require modification for this reason. Potters must react with flexibility and accommodation to nature's fickleness.

Frit glazes. These are glazes prepared substantially or entirely from frits. A frit is actually a glaze that has been melted into a liquid state, then cooled and ground into a fine powder (powdered glass). Each frit is identified by a number and manufacturer, such as "Ferro Frit 3304". Its composition is standard and uniform. Frit manufacturers helpfully make the compositions readily available.

Frits have the advantage of melting more trouble free than raw glazes, because they had been melted once before. Certain desirable oxides that are toxic in their raw states may be incorporated into frits. An example is lead. In frit form these oxides can be made safer for handling, and the fired glaze safe for use with food.

Some oxides are soluble in water, a great inconvenience because glazes are mixed with water for application. Melted into a frit, the soluble oxides are converted into an insoluble form. For these reasons frits are sometimes used as the dominant base ingredient around which glazes are composed. They may also appear as one of the minor ingredients in a glaze. Frits are considerably more expensive than most raw materials.

Glaze Composition

The dominant material.

Lead glazes. Lead has been, for more than a thousand years, a very much utilized and desirable glaze ingredient. The contributions it makes to a glaze are impressive, such as excellent transparency, brilliant luster and great compatibility with colorants. Lead helps glazes melt extremely well; it is a remarkably good flux. For these reasons, lead has been a popular glaze component. Unfortunately, lead can be toxic when it is handled in the raw or unfired state and when lead-glazed pottery is used with certain foods. There are methods to get around this great disadvantage, involving hygienic studio practices, adjustments to the empirical formula, and substitution of certain frits for raw lead. (See Chapter 19 for health and hygiene in the pottery studio.) Lead reacts to heat above cone 4 by vaporizing. Consequently, lead is not a factor in those glazes, and they may safely be considered free of lead.

Leadless glazes. The absence of lead in glazes fired below cone 4 is significant in that some other flux will have been substituted to do the work; hence, the title "leadless." Above cone 4 lead is virtually never included in glaze formulas, because, as noted above, it vaporizes above that temperature. Growing concern about lead as a health hazard has caused greater use of other good fluxes, such as: gerstley borate (colemanite), leadless frits, and nepheline syenite. Each of these makes a unique contribution to the glaze's chemistry. The precise effects on color, texture, and transparency are not the same when glazes in which these fluxes are used are compared with each other or any of them with lead glazes.

Alkaline glazes. The major alkalis, potassium and sodium, are used as fluxes in some glazes. Their effects on a glaze are in a distinctive shininess somewhat harder than that of lead and in the opportunities for characteristic colors. For example, copper oxide in a highly alkaline glaze results in a turquoise color, as opposed to the green of a lead glaze. Crackle patterns are encouraged much more successfully when the formulation is heavily alkaline. In medium- and high-fire wares, such as stonewares and porcelain, potassium and sodium dominate as flux ingredients. The alkalis are introduced largely in the form of feldspathic materials.

Slip glazes. A few low-fire clays, such as Albany clay, are unique in their property to melt successfully into glaze. These clays are applied to bisk pottery in the form of a slip (see Chapter 8).

Salt glaze. In this technique, some manner of salt (sodium chloride) is introduced into the kiln, usually when the fire is quite hot. The salt is vaporized by the high heat; the vapor permeates the kiln's atmosphere and forms a glaze coating where it comes into contact with pottery surfaces. The glaze is created from a fusion of sodium from the salt with silica from the clay. The knowledge of salt glazing goes back more than 500 years. It can be recognized by its telltale orange-peel texture. Salt glazing should not be attempted in electric kilns, for the heating elements are attacked by the fumes. Salting produces dangerous chlorine gas as a by-product, and the procedure should always be done out-of-doors in isolated areas.

Cone values

a. Low fire glazes: cones 015 to 05.

b. Medium fire glazes: cones 04 to 5.

c. High fire glazes: cones 6 to 15.

Most glazes have cone values that characterize their best melting condition, such as, "This is a cone 9 glaze." (See Chapter 13 for a full discussion of cones.) Some glazes melt satisfactorily over a range of cones. The empirical formula of a glaze can be adjusted so that it melts at a desired cone, in addition to any other of the potter's technical and visual requirements for that glaze.

Texture

a. Bright, shiny glazes.

b. Matt, or dull, buttery glazes.

Bright glazes have highly reflective surfaces. All transparent glazes are by their chemistry bright, but opaque glazes may be either shiny or matt. Matt describes a broad range of less reflective textures such as: buttery, soft satin, pitted, velvet, coarse, or sandy. Matt glazes are somewhat opaque because of their chemistry. Some glaze textures are neither highly reflective nor dull and are referred to as semi-matt glazes.

Transmission of light

a. Clear glazes.

b. Opaque glazes.

Clear or transparent glazes are "perfectly" melted glazes. They reveal with maximum clarity the surface quality of the clay beneath. Clear glazes are also bright in texture. Opacity in a glaze layer is created when some ingredient present remains in suspension and interferes with the transmission of light through the layer. These ingredients, often deliberately introduced into the glaze formula are called *opacifiers*. They include tin oxide and superpax. Color appears more vibrant in opaque glazes. Without color opaque glazes are white—examples are majolica or Bristol glazes. Many glazes fall between the two extremes and may be semi-opaque or translucent.

It has been seen that glazes can be complex affairs. The system of classification outlined above presents the basic qualities of glaze in a logical organization. It gives the potter a practical way to comprehend glaze. To the extent that the reader is alert to the range and interdependence of these fundamental factors, he or she will be supported and amply rewarded in creative and sensitive glazing.

PURCHASED GLAZES

Probably the most convenient way to obtain glazes is to buy them ready prepared. They are sold by most pottery suppliers in liquid and dry measure. *Commercial glazes* come in a wide variety of colors, textures, and opacities; lead or leadless; formulated to melt properly at specified cone ranges. Glazes purchased in powder form need only be mixed with water into a creamy liquid. Liquid glazes may be used straight from the container. These glazes keep well indefinitely. Should the glaze dry, add some water and mix well to the desired viscosity.

It is important to have glazes that will melt suitably at a temperature the same as or below that of the maturing cone of the clay body. For example, clay destined to fire to cone 04 should have a glaze whose cone value is cone 04 or lower.

Beginning potters who purchase prepared clays and glazes might be well advised to get their materials from the same supplier to increase the likelihood that the glazes are compatible with the clays. There is less chance for glaze faults to develop in firing and cooling.

GLAZE RECIPES

Many potters prefer to mix their own glazes from scratch. For some individuals, there is great personal satisfaction derived from having a hand in all phases of pottery making. Potters also have a sense of greater discretion and more precise control over the nature of their glazing when they mix their own. Even though the range of commercial glazes is broad enough, individualized effects and visual idiosyncracies may be more successfully developed by the potter's own efforts. Pound for pound and gallon for gallon, studio-prepared glazes can be more economical.

Glaze recipes abound. The best source of good recipes is another potter. Most are forthcoming in sharing information and swapping favorite recipes. One important virtue of this source is that the glazes will have been tested in actual use. Books, too, are a good source, and hardly a pottery book is published without the inclusion of some recipes. In recent years, a few books have appeared which consist of nothing but recipes. The pottery magazines often print recipes.

No matter the source, put any glaze recipe to the test by weighing, mixing, applying, and firing a sample on a test piece. (This procedure is described in great detail later in this chapter.) A cautious potter might even go further, by mixing a larger batch of the glaze for a test on a pot or two, before committing material, effort, and time to quantities of work.

Glaze recipes that have been around for a while have a mystifying tendency to produce different results when they are resurrected. This is equally true for glazes adapted from books and magazines. Sometimes the raw materials are no longer the same as they were in time gone by. Perhaps firing conditions have changed, the clay is not the same as before, or the glaze is applied differently. These variables, or some unknown element, sometimes dramatically influence the glaze mixture, even though the recipe on paper is as before.

The following sampling of glaze recipes might be considered by the reader as "base recipes." A *base glaze* is one that is fundamental in its qualities. Its destiny is to be altered, for transformation in the properties of color, opacity, and texture.

CONE 06		CONE 04	
Leadless Clear		**Good Clear Bright**	
gerstley borate	13.3	lead monosilicate	237.6
magnesium carbonate	10.8	potash feldspar	35.5
nepheline syenite	29.7	soda feldspar	5.1
kaolin	8.5	EPK kaolin	45.6
flint	37.7	flint	13.4
	100.0		337.2

CONE 4		CONE 9	
Good Clear Bright		**Good Clear Bright**	
gerstley borate	48	potash feldspar	178.6
flint	30	whiting	70.0
kaolin	27	kaolin	28.6
	105	flint	110.6
			387.8

ALTERING GLAZES

The base glazes shown above melt into shiny, clear coatings. They completely reveal the color and texture of the clay underneath. The chemistry of each, however, varies. As a consequence, the same clay seen through each clear glaze is slightly modified. In addition to their clarity, these glazes are extremely reflective, very bright. There is little else to recommend them visually. On some pots or over some decorations these base glazes might function well; other pottery forms seem to clamor for less transparency, different textures, and a wider range of color. Fortunately, these base glazes can easily be transformed into interesting design instruments. Most other glaze recipes can be altered in the same manner. Percentage additions of selected oxides to the glaze change the fired appearance of the glaze properties, that is, the *texture*, *color*, and *opacity*.

These dimensions may be altered singly or in combination. When more than one oxide is added to a glaze, their interaction furnishes additional visual change. Blue colorant in an opaque bright glaze does not give exactly the same blue as if it were in a matt glaze, even though the proportion of blue is identical in both examples. The chemical environment of a glaze is a decided influence on how an added oxide affects the fired state. Experience also teaches us that when we add more than one oxide to a glaze mixture, we do well to reduce the percentage of each. Otherwise, they tend to overcome each other and saturate the glaze. This risk arises when colorant oxides are being considered.

A few ingredients comprise the list for altering opacity and texture. Colorant oxides are more numerous; these are given in Chapter 9. Percentages suggested here are within a range from a slight influence to a much stronger effect. The reader may choose to use less or more than the suggested percentages.

CHANGE FROM A CLEAR GLAZE TO OPACITY

superpax, opax, zircopax	add from 5 to 15%
tin oxide	add from 3 to 8%
rutile	add from 3 to 15%

Tin oxide and the zirconium compounds superpax, opax, and zircopax are the opacifiers most often chosen. Tin oxide is probably the preferred material for the quality of its opacity and its effect on color, but the others may often be substituted for reasons of economy. All produce a white or slightly off-white opacity. Their effect on texture is imperceptible; the brightness of the glaze remains virtually unchanged. Rutile is a potent opacifier, but it also conveys a streaked or mottled mattness and changes color to sand or buff. Nevertheless, rutile is a rich provider of surface quality and is a favorite among potters.

CHANGE FROM A BRIGHT GLAZE TO MATTNESS

barium carbonate	add from 2 to 8%
kaolin	add from 5 to 20%
rutile	add from 3 to 15%

Barium carbonate and kaolin each result in distinctive velvet, matt textures. Sometimes a glaze is named after the characterizing oxide, such as a "barium matt." Rutile produces a somewhat streaked or mottled matt, sandy or buff in color. A by-product of all matt alterations is a measure of opacity. Some matt glazes cover thoroughly, and others are semi-opaque, but there is no such thing as a completely clear matt glaze.

TO CHANGE A GLAZE'S COLOR

Color in glazes is always affected by the chemistry of the glaze to which the colorant has been added. For example, the identical percentage of cobalt in two different base glazes results in two different blues.

Color in a clear glaze is strongly modified by the color of the clay beneath the glaze. A blue transparent glaze over red clay does not appear blue, but over a white clay body it is clearly blue.

Color in a bright (shiny) glaze seems more brilliant than the same oxide in a matt glaze. Color in high-fire glazes appears more subtle and subdued than the same material in low-fire glazes. Firing conditions, whether electric, oxidation, or reduction, have profound effects on all the properties of color. A table of colorant oxides is given on page 82.

Arithmetic for Changing Glazes

Calculations for alterations of opacity, texture, and color in a glaze are easy. The following example should serve for most kinds of alterations:

1. The separate ingredients in a glaze recipe, when added together, give us the *batch weight.* There are 453 grams to the pound, and approximately 2000 grams make a gallon of glaze after water is added.

2. Determine the quantity of glaze you need. Should more than a single batch weight be wanted, multiply each recipe ingredient by a common number. For example, if you require four times as much glaze, multiply each material in the recipe by four. Conversely, if a smaller quantity were desired, as in the preparation of a glaze-test sample, divide each of the ingredients in the recipe by an appropriate common denominator.

3. Decide which additives will be used and the percentage of each. For example, you may want to make the glaze more matt, increase its opacity, and color it green. The additives for these alterations could be barium carbonate 5%, superpax 6%, and copper carbonate 2%.

4. Convert each percentage to its decimal equivalent: i.e., barium carbonate .05, superpax .06, and copper carbonate .02.

5. Add all the ingredients for your new, adjusted batch weight.

6. Multiply in turn the new batch weight by the decimal equivalents of each additive.

7. The product of each multiplication is the amount of that material to be weighed and added to the glaze, at the time the glaze is prepared.

A helpful description for how to weigh, mix, and prepare glazes follows.

PREPARING A GLAZE

TOOLS AND EQUIPMENT NEEDED

1. Laboratory scale capable of weighing in grams.

2. Spatula, large spoon, or scoop for handling dry-glaze ingredients.

3. 40- or 50-mesh sieve, or coarse meshed screen to sift the combined ingredients in their dry condition. In the absence of a pottery sieve, a suitable substitute can be improvised. Staple a one-foot-square piece of aluminum window screen to a wooden frame.

4. Mortar and pestle, 32-ounce size, or a similar large size for grinding the glaze mixture with water.

5. Some clean newspaper.

6. A sponge and some soft artist's brushes.

7. Containers for storing prepared glazes. They should be rust proof, have lids, and should have mouths large enough for the hand to fit through. Ideal are the gallon-size plastic jars used by restaurants.

WEIGHING THE INGREDIENTS

1. The metric scale, conveniently in place on a work counter and dusted clean of residue from prior use, should first be accurately balanced. The purpose is to bring the scale pointer to a zero reading to which it returns after each material is weighed.

2. Weigh in turn the full amount of each ingredient. Most pottery scales have some arrangement, such as a scoop or deep pan, in which the glaze material is placed for weighing. Should the scale's scoop prove too small for the quantity being weighed, do it in two batches.

3. Deposit each weighed material separately on a clean newspaper sheet. Write the name next to each pile of powder. This simple housekeeping device acts as a check, to ensure that all the recipe has been accounted for and that none of the ingredients are weighed twice.

4. After all the materials have been weighed, including the alteration oxides (color, texture, opacity), combine them in one heap.

MIXING THE GLAZE

1. Pour the entire dry glaze batch into a clean screen or laboratory sieve, held over an opened, double-layer newspaper sheet. Sift it through the screen four or five times. Sifting is an excellent method to thoroughly combine and mix together assorted dry ingredients. (See Chapter 19 on health precautions.)

2. Pour enough water into a mortar to make a creamy liquid with the dry materials. Generally, a ratio of about one cup of water for each 300 grams of dry ingredients seems to give good results. The amount of water can be adjusted later, but it is probably easier in the long run if there is a bit less liquid to start with, to avoid a watery glaze.

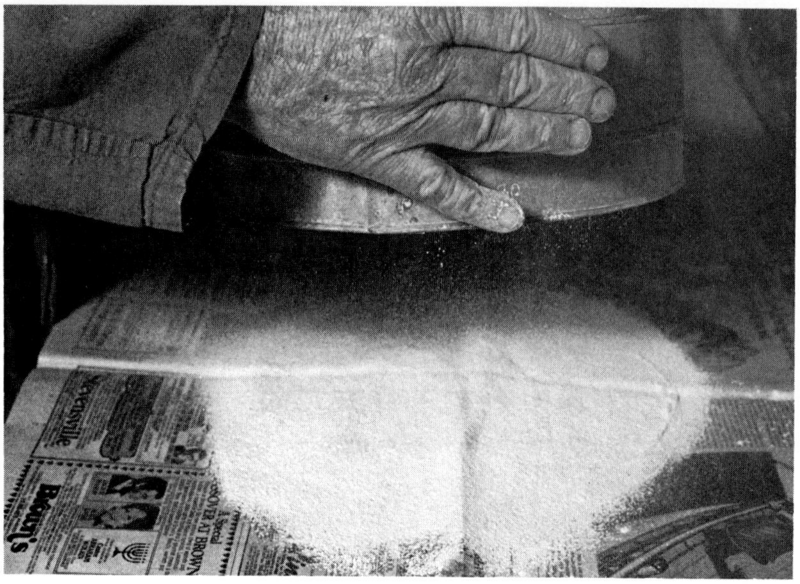

14-1 *Sifting a batch of glaze.*

14-2 *Grinding glaze with mortar and pestle.*

3. Grind the glaze with the pestle. Do a thorough job; some "elbow grease" helps. The glaze should be a smooth, creamy, or syrupy liquid, without lumps or separations. If the mixture is too viscous, add some water and grind the glaze smooth. If too much water has been introduced, allow the glaze to sit undisturbed, until materials settle to the bottom and clear water rises to the top. The excess water may now be decanted.

STORING GLAZE

1. Pour the glaze batch into a clean storage container, one large enough for the entire amount. Tightly covered, the glaze keeps indefinitely. Label the container for identification.

2. In time the glaze may lose some of its water, through use and evaporation. More water may be added to bring the glaze to a proper consistency. Stir by hand or with a paddle until smooth.

3. After use, lumps of hardened glaze matter may appear in the glaze mixture. Attempt to break these up by hand. It may be wise to put the liquid glaze once again through a screen, to separate the lumpy substance. This is best accomplished with the same kind of screen or sieve, but one that has first been made wet, to minimize clogging of the wire mesh.

These procedures serve the purpose of preparing small test samples, as well as large glaze batches. Time might be spent more productively if several different tests of the same oxide were attempted at one time, for example, the base glaze altered in a series of trials: 2%, 4%, 6%, 8%.

Once having determined good opacity and texture from these fired results, a series of color trials may be conducted. Colorant oxides may be added singly or in combination. For example, if the blue of cobalt appears too raw, it can be modified by the addition of iron by using the recipe plus cobalt oxide at .05% and red iron oxide at 2%. For a blue-green use cobalt oxide at .025% plus copper oxide at 1%. For fun, make a number of color-test combinations by blind guess. These colorant trials may be more productive if a range of percentages is used, progressing from minimal amounts to larger quantities.

Beginning potters may find themselves spending too much time on these experiments. It is intriguing to formulate and mix the powders and exhilarating to observe the range of glaze qualities that are obtained when the trial pieces emerge from the kiln. As in all art, however, one must set priorities; and in creative pottery work the first considerations should be sensitive form and surface. A sensible glaze testing program will serve the practical needs of the potter.

Obviously, glaze is not intended to be ingested, even though most of the ingredients are neutral and harmless. Some of the potter's materials are toxic and should be dealt with cautiously. As a rule, avoid handling any glaze materials or the mixed glaze if there are open cuts or wounds in the skin, and do not eat, drink, or smoke while glazing. Some potters habitually wear dust masks while weighing and mixing glaze and other dry pottery materials, but they are ineffectual unless certified for this kind of dust. The studio should be well ventilated and kept clean. (Safety and hygiene in the potter's studio is more completely described in Chapter 19.)

GLAZE APPLICATION

In days gone by much glazing was done by dusting the dry glaze compound over the pot. Glazing can also be effected during the firing process by salt glazing, introducing a quantity of salt into the red hot kiln atmosphere. In wood-burning kilns, the falling ash residue deposited on the pottery may be transformed into a fused coating; this is an *ash glaze.* However, most glazing is accomplished in a practical manner by coating bisk pottery with a coating of liquid glaze. The glaze coating dries, following which the pot is put through a second firing to melt the glaze.

Any means employed to get a proper coating on the bisk pot will do. Potters have practiced glaze application techniques which might seem to us strange. But there are no objective or uniform standards for application. The critical element, no matter by which method the glaze liquid has been applied, is *how thick a coating remains on the pot's surface.* This is the primary determinant for how the glaze will melt. The *thickness dimension* is the location of the glaze oxides, where they fuse into a good glass with desired visual and technical properties. The potter acquires skill in glazing to the extent that he or she can control thickness of application.

Many glazes seem to melt well over a range of thicknesses, but their visual properties vary with the thickness of the coating. It is prudent to try a new glaze to determine in which ways thickness affects its technical and visual characteristics. The beginning potter might be well guided by the following general suggestions: *matt glazes may be applied thickest, clear glazes thinnest, and bright opaque glazes somewhere in between.* The average thickness in this measure is somewhere in the neighborhood of shirt cardboard.

Some potters practice glazing bone-dry greenware, but they are few. The technique is referred to as *single firing.* There is difficulty in handling bone-dry pots while glazing, as they are bound to be quite fragile and might also soften under the flood of liquid glaze. But the major problems arise during firing. The transformation of greenware into bisk has enough problems of its own, without adding the problem of a glaze melting simultaneously in single firing. Extraordinary precautions are taken, and special skills and routines must be developed. For these reasons, the prudent practice rather than the exception has always been twice firing, bisk fire first followed by a separate glaze fire.

Probably the major glazing procedures used today are dipping, pouring, spraying, and brushing. They are all good methods to get an appropriate thickness of glaze on pottery. A description of each follows, but in no particular order of difficulty, preference, or other significance.

Dipping

Dipping is a simple and effective means to cover the entire piece all at once or to coat uniformly just one portion of the pot. It is much like dipping an apple into jelly or caramel. Of all the glazing techniques, dipping requires the largest quantity of glaze on hand, although it is, at the same time, the least wasteful of glaze.

Have ready glaze in a container large enough for the pot to be dipped in, without the glaze spilling over the sides of the container.

First, the bisk should be prepared for glazing. Clean the pot thoroughly by washing it under running water or by sponging it with clean water. All the dust from the studio is washed away, and at the same time the bisk becomes wet. The longer the wetting the more the bisk is soaked. Observe that water is absorbed by the porous bisk, as if it were a blotter. Leave no trace of moisture on the surface; if necessary wipe the pot dry with a clean cloth or paper towel. Avoid handling the piece except as absolutely necessary; keep the surfaces free of dust and skin oil.

Probably the most important influence on how a glaze turns out is the thickness of the applied coating. Thickness of glaze coating is a function of the relationship among these factors:

1. The viscosity of the glaze (how watery the glaze is).
2. The porosity of the bisk (how much the bisk is soaked in water).
3. How long the bisk pot is kept in contact with the glaze.

If the glaze mixture is more viscous the glaze coating will be thicker.

If the bisk has been washed longer and is more saturated, bisk porosity is diminished, and the glaze coating will be thinner.

The longer the pot is kept dipped in the glaze bath, the thicker the coating.

The potter learns to arrange the three variables in the effort to apply a suitable glaze thickness. One condition is balanced against the others. For example, if the glaze liquid is watery, then the bisk need not be soaked as much, and the pot may be dipped for a longer time.

The glaze bath must always be kept uniform in consistency; if it rests for 30 seconds or more, the minerals in the glaze will begin to settle.

Hold the piece at the footing, rim facing down. Slowly lower the pot straight down into the glaze bath until it is entirely submerged up to the finger tips. Then slowly draw the piece out of the glaze. The entire exterior up to the finger tips has been given a uniform coating, but air pressure within the vessel has prevented glaze from reaching up inside. The interior should be glazed first by pouring, or it will remain unglazed.

Dipping is a good method to apply glaze to both the interior and exterior, all at once. Hold the piece horizontally, between two fingers, one at the rim and one at the footring. Lower the pot sideways into the glaze bath, submerging it half the distance of its circumference. Slowly lift it from the glaze, while tilting it forward for the glaze to run out of the interior. The pot must be held until the glaze is dry to the touch; this does not take very long. Reverse the pot's position, and dip the unglazed, second half until the glaze layers make contact. The second dipping might even be in a different glaze.

When two differing glaze qualities are side by side on the same pot, the potter must consider what they do for the pot. Are the glaze differences resolved in such a manner as to strengthen the

14-3 *Dipping.*

14-4 *Fired dipped glaze application.*

pot's total visual impact? Or are the contrasts so strong that they seem to destroy the piece, to bisect it?

The piece may also be held and dipped at oblique angles for a variety of irregularly shaped coatings. If desired, layers may be overlapped or superimposed, by dipping once again into the same or different glazes. There is no limit as to the number of coatings of glaze that may be piled on top of each other, as long as the total thickness melts well. Permit the first coat to dry slightly before immersing the pot into a second glaze bath. But if the first coat has been allowed to become too dry, there is the risk that it will pull away from the pot when a new wet layer comes over it. Moist or wet glaze coatings should never be touched; the finger mark becomes a permanent record within the melted glaze.

There is no practical means to measure the coating thickness until the glaze dries. Pierce through the dry glaze layer down to the bisk surface with a needle or knife point. The thickness dimension becomes readily visible. A glaze that is too thick must be removed. A thin coating may be carefully reglazed for added thickness, but a beginner might find it simpler to remove the whole thing. First scrape the dry glaze with a knife blade or scraper, catching the scraps for salvage. The remaining glaze can be wiped clean with a moist sponge and the pot reglazed.

Dipping rims is easy and convenient, even when the rest of the piece has been glazed by some other method. The rims might be dipped into a different glaze from that of the rest of the pot.

Feet take special preparation. Most potters prefer that the feet be completely bare of glaze, to avoid the risk of glaze dripping down over the base to fuse it to the kiln shelf. Wipe the footring clean with a sponge, and a little way up the wall, perhaps 1/16″ to 1/8″. Even though stilts or pins might be used to support the pots during firing, it is prudent to have footrings clean for the stilt's points to rest against. Take all precautions to avoid having your pot fused to a shelf.

Pouring

Pouring is another widely used glazing technique. Glaze held in a cup or ladle is carefully poured or spilled over, around, and into a pot. What does not stick to the bisk piece runs off into a catch basin placed underneath for that purpose. *Pouring is almost always how the insides of pots are glazed.* It is also a good method to cover entirely the exterior of a pot or to contribute dynamic effects of spill, pour, and glaze flow. The glaze that runs off during pouring may be manipulated for a deliberate, energetic pattern of overlays in translucency, opacity, and color. However, the potter can easily overdo spill and dribble, creating an effect so contrived, fluid, and busy as to destroy the significance of the pot's form.

Prepare the bisk in the same manner as described for dipping. The thickness of the glaze coat is instrumental in how well it melts and cools. Control of thickness is accomplished the same as in dipping: by balancing the wetness of the glaze liquid, the porosity of the bisk, and the length of time the glaze is permitted to flow over the piece.

The interior of a vessel is most easily and effectively glazed by pouring and generally is the part of a pot that is glazed first. Holding the piece with one hand, fill it quickly to about the two-thirds mark. Immediately turn the pot around in the hands, tilting it slowly in rotating and from side to side so that the glaze flows around making contact with the entire inner surface, covering up to the rim line but

14-5 *Wipe the footring clean with a sponge.*

14-6–14-8 *POURING THE INSIDE OF A POT.*

14-6 *Fill the pot about two thirds full of glaze.*

14-7 *Swill the glaze around to cover the interior surface.*

14-8 *Pour out the excess glaze.*

not spilling over. Turn the piece over at once for the excess glaze to flow out, passing over the pot's rim into the catch basin below, revolving the pot slowly for the flow to be distributed uniformly all around. Hold the pot upside down until the glaze is no longer runny; it takes a few moments. Pouring is done quickly; the entire inside of a pot takes less than a minute to glaze. The movements of the potter, while swift, must also be smooth and uninterrupted for maximum control and to avoid glaze skips, pile ups, splashes, and drools.

Pouring is an excellent means to give the piece a uniform coating of glaze and is even more adaptable for the potter to dynamically express movement and interchange in line, shape, color, texture, and translucency. Glazes may be overlapped and superimposed, poured and spilled coatings in single and multiple layers. One or several different glazes may be used on the one pot. Some glazes are flexible and react beautifully to variations in thickness, perhaps translucent and a bit shiny where thin and more opaque and more matt where thicker.

Glazes may be poured over dipping and dipped over pouring. A miraculous layering of translucencies may result which magically brings surface and form into an integrated whole.

Pouring over exterior surfaces is done after the inside has been glazed. Place the pot upside down, resting on two sticks over a catch basin. Small pots can be held in one hand between thumb and forefinger, one at the footring and the other at the rim. Pour the glaze with a cup or ladle. Hold the cup full of glaze, its lip at the pot's footring level. Circle the pot, simultaneously tipping the cup and allowing a stream of glaze to flow steadily around the pot's perimeter. Glaze issues from the moving cup, around and down from the footring, gradually covering the pot's exterior surface. With some practice a good coating of uniform thickness can be easily managed. This is a particularly effective and fast technique for overall glazing of large work. Glaze run-off is salvaged in the catch basin.

Pouring may also be used for other than a uniform coating. One or more glazes can be selectively splashed, spilled, dribbled, and poured over portions of the pottery surface. Poured freely, the glaze falls into expressive and dynamic shapes, while at the same time the streams of glaze create new contours or negative shapes. The effect of glaze poured in streams is linear movement and rhythm. While the results may at times seem overly casual or accidental, with practice and care a remarkable degree of control and predictability can be achieved.

As in dipping, risks arise when glazes are poured over previous layers. The coat underneath should be permitted to dry beyond runniness but not become too dry. Some moisture should remain in the

14-9 *Pouring glaze over the exterior of a small pot.*

14-10 *Pouring the exterior of a large pot. Support the pot by two sticks over a basin.*

layer below, before another coat is poured over. If too dry, the glaze coating underneath will pull away from the pot when a wet glaze runs over it. Beginning glazers must always be cautioned never to touch moist or wet glaze, for the evidence would be held in perpetuity!

The rims of pots whose glaze has been poured require special attention. The rim is apt to be ragged from the pouring process. Scrape the rim clean with knife, and then sponge it bare. The clean rim may then be dipped in the same or a different glaze, or glaze may be carefully brushed around the rim. In either case, it is extremely important to match the thickness of glaze at the rim with that on the pot's walls below. Clean footrings thoroughly, as described in the previous section on dipping.

Brushing Glaze

Dipping and pouring are the glaze-application techniques preferred by most experienced potters. Applying glaze with a brush is a more limited technique. But children, and those whose fingers and undeveloped motor skills cannot cope with the more complex hand and arm movements of pouring and dipping, are better off using a brush. Some pots, particularly sculptural forms, are so complicated in the intricacies of their concave and convex relationships that glaze brushing seems to be the most practical method. The piece may be just too large to handle any other way. At times, small areas or spots may have been inadvertently skipped and left bare by pouring and dipping. These portions can be easily filled in with a brush. Brushes are also used to paint decorative glaze lines and linear shapes over bisk and on unfired glaze coatings.

Flat brushes work best, soft bristled and at least one inch wide. Prepare bisk in the standard way by thorough cleaning. Brushed glaze builds up more successfully when the bisk is slightly porous, not too damp. Paint the first coat in parallel strokes, covering the area desired. Follow that by another brushing of parallel strokes, but across the first layer. This sequence of layers should be continued until the proper thickness has been applied.

The major disadvantage in glaze brushing is the difficulty in achieving a flowing, uniform coverage, one that is free of annoying surface flaws such as brush strokes, splotches, and thin and thick portions. These visual disruptions show up in the melted glaze. However, brushing glaze does not require nearly as large a quantity of glaze on hand as the other techniques and is probably the least wasteful of material.

Spraying

Spraying glaze is an excellent method to apply a perfectly uniform coating. It can also be a convenient and practical means to put several layers of glaze over each other. Colorants are easily sprayed on bisk and glazes. Perhaps the great drawback is the risk that the uniformity and convenience of sprayed glaze may express more the mechanics of the process than the potter's human creativity and sensitivity.

Certain tools and equipment are required for successful and safe spraying:

1. Spray gun. The instrument in which compressed air atomizes the glaze, to be forced out as

14-11 *A spray booth.*

American Art Clay Company, Inc., Indianapolis, IN.

a mist or spray through a nozzle in the gun. The potter holds the gun in one hand, aiming and controlling the spray.

2. Spray booth. A cabinet open on one side in which the piece is placed for spraying. The purposes of the spray booth are to capture all the excess spray and to eliminate the mist from the studio proper. Much glaze lands on the spray booth's walls, where it can easily be removed and reused.

3. Exhaust system. A powerful ventilating fan and duct system that sucks the glaze-laden air from the spray booth and conducts it away from the studio.

4. Compressed air. The propelling force for the spray, generated by a local air compressor or a central system for the entire building. Small, inexpensive air compressors are available for the studio. The air flows under pressure, through tubing, into the spray gun, where it breaks the glaze up into tiny glaze-laden droplets or mist and forces it through the nozzle as spray. The pressure of the air can be controlled by a valve.

Obviously these components add up to a sizable investment and a large share of studio space. It is unhealthy to attempt to spray glaze without an effective exhaust system for the spray booth. Toxic materials are present in some glazes and silicates are contained by all glazes. They would permeate the studio atmosphere, presenting a health risk to all.

Although spraying is a good way to get a uniform coating on pot exteriors, it is not an effective way to glaze interiors. Except for flat or very shallow vessels, such as plates and platters, the stream of mist does not accumulate uniformly in deep volumes. Pouring is a better method to glaze interior surfaces. Do this a day or two earlier than the spray application, for the mist builds up more readily on dry, rather than moist, bisk. Glaze globules adhere to the pot's surface because the porous bisk absorbs water from each droplet.

14-12 *A sprayed glaze accumulates on the pot as a fluffy, powdery layer.*

14-13 *The consistency of a sprayed layer (right) compared with that of a solid poured or dipped layer (left).*

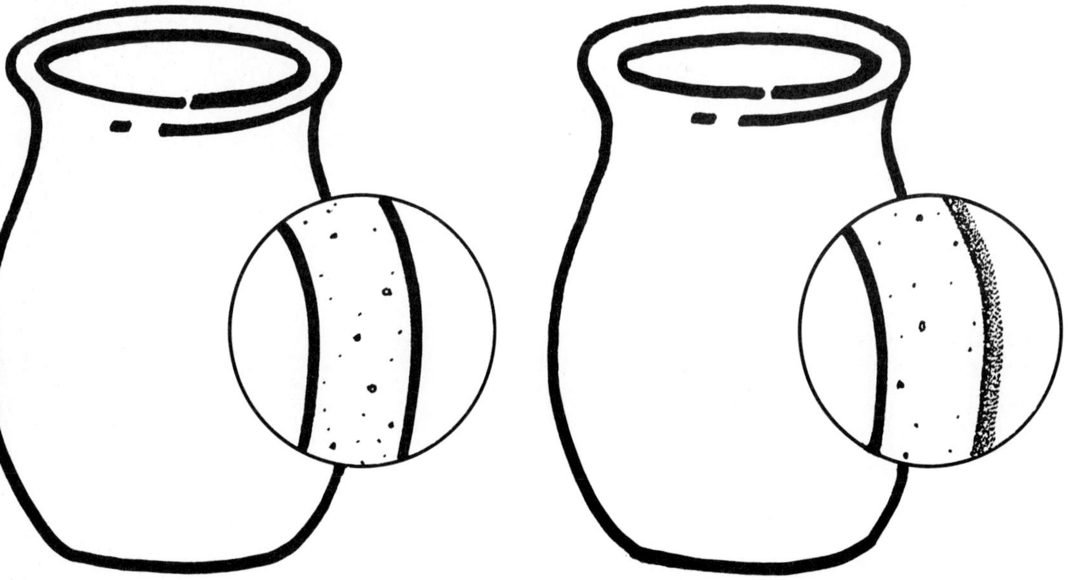

Place the pot upside down on a banding wheel or bench whirler that is inside the spray booth. First spray the bottom walls, the base, and inside the footring. Operate the gun in short bursts, to avoid a build up of wetness or runny drools. Hold the gun a foot or so distant from the pot for the mist to be uniformly distributed over the surface, rather than concentrated unevenly and at random. Properly sprayed, the mist accumulates on the pot as a fluffy powder, its thickness permeated with air pores, like a dry sponge. Compensate for the aerated coating by applying about 50% more glaze thickness than if you were pouring or dipping. The dimension of the glaze layer may be easily determined by piercing through to the hard bisk with a needle point or knife edge.

When the bottom walls have been adequately sprayed, clean the footring by scraping and sponging. Turn the piece right side up to rest on its foot, and continue spraying the exterior walls until completed. Test for thickness. Generally, the rim gets a good application at the same time. If there appears to be a risk that some of the spray might contaminate the pot's interior, cover the mouth with a piece of weighted cardboard or a jar lid.

Spraying is an efficient means to apply layer upon layer of glaze. The separately applied coatings fuse into one mass in the heat of the kiln. The diverse oxide compositions often react with each other to effect exciting dynamics of color, texture, and opacity. But make tests of these multi-layered compositions before committing good work, for it could turn out poorly.

The spray gun may also be used much like an air brush to spray colorant oxides on bisk, or dry, unfired glaze coatings. Make a mixture of the colorant with water, and spray it on the pot. Colors may bleed into each other. Covering parts of the pottery surface with stencils and wax will also yield interesting interchanges of positive and negative shapes.

SUMMARY

The danger exists that glazes may be chosen and applied for purposes pertaining solely to the glaze, apart from any reference to the compositional needs of the total work. The effect might diminish the overall coherence of the piece, in direct proportion to the splendor of the glaze's virtuosity. Sheer delight in a glaze procedure may dominate and visually destroy a piece. Strong and meaningful pots may be made mediocre by unharmonious glazing.

In the best sense, glaze transforms the pot, as if by magic, from sterile bisk into vital existence. Strange as it may seem, merely a thin coating of glass melted around the clay volume brings forth a new, whole being, where before there was an incomplete thought. The glaze serves the form, and also gives form conviction. We perceive that the glaze appears as natural to its clay form as skin to its human form.

15 GLAZE AND FIRING DEFECTS

"While it may chance that body and glaze and fire are so adjusted that faults do not develop, this state of things is rare—it is always possible that an occasional trouble may arise." Charles F. Binns wrote this in the first comprehensive American pottery text book (1910), and it still holds true. It may appear to be an overly cautious statement, but a potter who is not careful will find "faultless" glazes indeed rare. No matter how well we observe accuracy, order, and correct methods, defects do crop up. Nearly all the causes can be identified, and most flaws may be remedied or avoided. But once in a while the "mystery" occurs, a firing effect with no apparent cause, no ready explanation. Fortunately, the same mysteries rarely return to haunt the potter. The more commonly occurring defects and their causes are described below, together with some of the remedies the potter might attempt to effect a cure.

Blistering or Bubbling Single blisters or clusters of them are seen in the cooled glaze. Gases emanating from glaze materials, from between the pot and the glaze coating, or from the clay body have become trapped within the melted glaze. Nearly always when glaze melts, there are these gaseous emissions. If we peer through the peephole after red heat first appears and close to the time the warning cone is to deform, glazes will appear to be erupting like boiling oatmeal or lava. In the normal run of things, the gases all work their way through the viscous, molten glaze layer into the kiln atmosphere. The glaze then settles down into a healed surface in time for the cone to melt. But if the glaze or body ingredients contain an excess of gas-producing sulphates or carbonates, and if the glaze coating is too thick, the ideal conditions exist for blistering. Be careful not to glaze too thickly, particularly when a shiny glaze is involved.

The potter can try to cure a blistered glaze or at least to help make it better. With a knife blade scrape the bubbles, until they are broken. Do not rub them with the fingers, as this will risk scraped and lacerated skin. Fire the piece once again. A soaking fire at the end should be a big help (see Chapter 12).

When blistering occurs frequently, the cause may lie in the clay body. Suspect that it contains too much sulphate. If it is practical, the potter could then add about 1% of barium carbonate to the clay during the mixing stage. If that is impractical, the clay should be changed.

Pinholes A sharp point appears to have picked away at the molten glaze leaving telltale pin pricks over the surface. These tiny holes are actually the aftermath of blisters. The molten glaze was on its way to healing over the blisters, but the kiln, upon being turned off, cooled too rapidly.

Rub some glaze into the pin holes with a finger tip or small brush to fill the holes. Fire the pots once again. A soaking fire at the end should help.

If the problem of pin holes persists, the cause may be a kiln that, not adequately insulated, loses heat too rapidly when it is turned off. An attempt should be made to increase heat retention. Fill any cracks with fire clay mortar, or cover the top of the

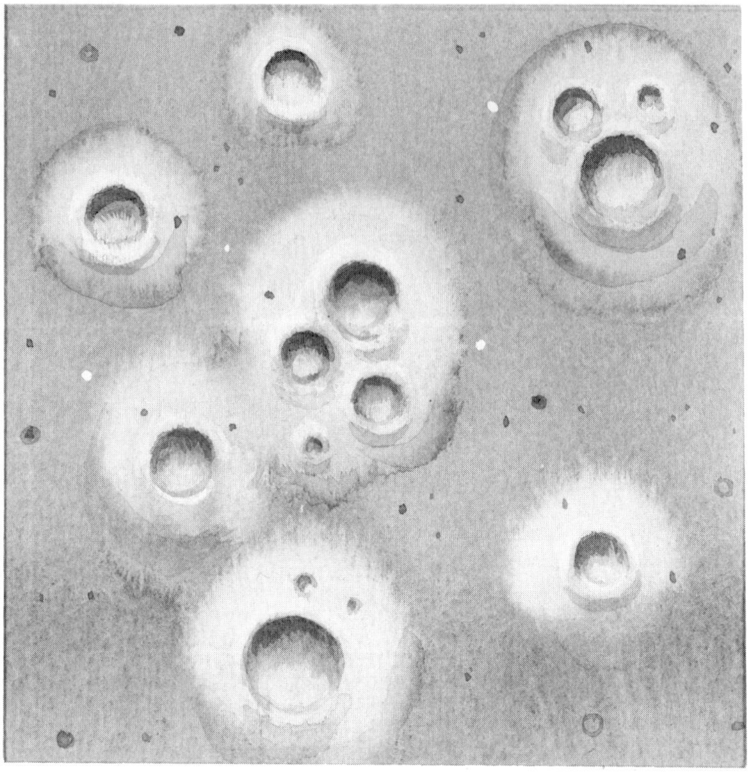

15-1 *Blisters.*

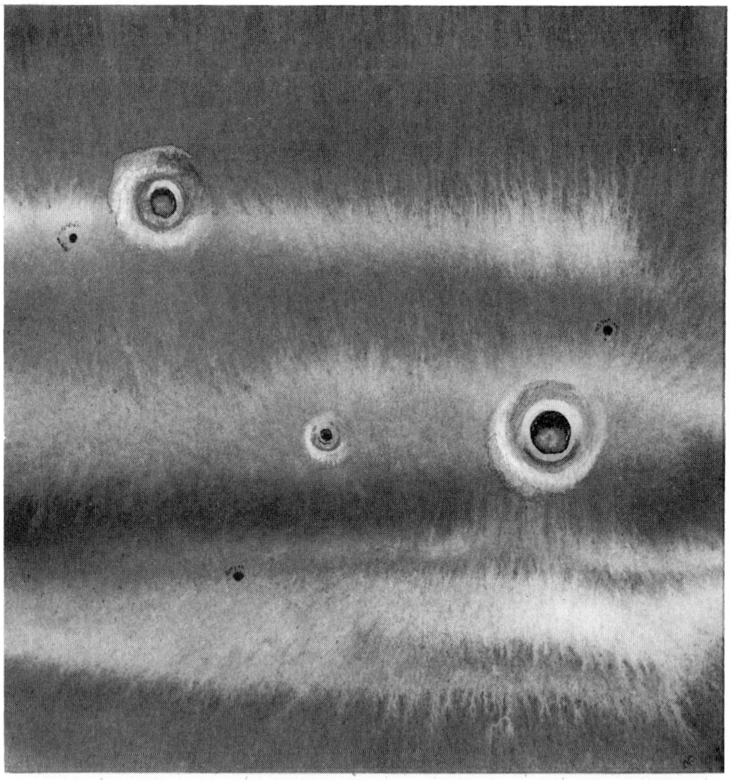

15-2 *Pinholes.*

kiln with a ceramic fiber blanket or aluminum foil and vermiculite. It may help to leave the kiln on a lower setting for a brief period after the cones have fallen and to "soak" the kiln regularly at the end of each firing (see Chapter 13).

Crazing A network of cracks appears in the cooled glaze. Crazing is a glaze defect, although sometimes it is sought for the enhancement of the glaze's appearance. Crazing thus induced for decorative purposes is renamed a *crackle glaze.* Nevertheless, craze or crackle, it is flawed because it is broken through, no longer waterproof or sanitary. Left alone, the crazed lines will, in time, darken with dirt.

Crazing is complex. Diverse conditions may be

15-3 *Crazing.*

responsible for a glaze pulling apart around the volume of a pot. Each cause has its own remedy.

Some glazes react adversely when the pots are taken from the kiln still too warm. Thermal shock causes the glaze to crack. Try leaving your work in the kiln until it reaches near room temperature.

Delayed crazing appears after weeks or months. The indication is that the bisk is underfired and remains soft and absorbent. Delayed crazing is stimulated by the clay's absorption of water, such as from dish washing or high atmospheric humidity. The bisk expands just enough to break the hard glaze coating around it. The obvious remedy is to fire your work till the clay body is mature.

Each clay and glaze reacts to temperature change by expanding and contracting. Crazing may result when the pot and its glaze are, in these terms, incompatible. Potters refer to this inescapable phenomenon of glaze adjustment to body, as the *glaze fit*, how the glaze fits or suits the clay body in terms of their relative expansions and contractions. If firing and cooling are well under control and crazing still occurs, the potter might attempt a cure by adding to *either* the clay or the glaze a few percent of potter's flint (silica). The flint is added at the time the mixture is being prepared (see Chapters 5 and 14). This cure is practical only when studio-prepared materials are in use. When potters purchase their clays and glazes commercially prepared, it would be prudent to get them from the same supplier to make it more likely that they had been formulated for compatible fit.

But for the glaze that has already crazed, there is no cure: It remains crazed.

Some pots may be beautifully decorated with a web of broken connections that appears like tracery work beneath the transparent glaze. Of course, the web is no more or less than crazing. But the pattern of cracks has been manipulated in spacing, direction, and intensity to integrate well with the form.

Potters can work with glaze fit, as they first did a thousand years ago in China, to develop lovely crackle glazes. They can adjust body and glaze ingredients to take advantage of the relationship between expansion and contraction and by modifying glaze thickness. A colorant is often rubbed into the cracked lines to heighten contrast and show the pattern of tracery more dramatically. Colored inks, shoe polish, and other non-ceramic pigments may also be used. Crazed and crackled glazes are really not functional for the serving of food, although it is technically feasible to apply and fire a clear glaze over the crazed glaze, which effectively seals the defect.

Crawling When glazes, upon melting, roll up into clumps or islands separated from each other, the defect is called *crawling*. The clay is exposed and bare between the gatherings and clumps of glaze. Usually, this is an unsightly mess. Crawling occurs more frequently with matt glazes than with those that flow more. Sometime crawling results when the bisk is much underfired or very dry and dusty

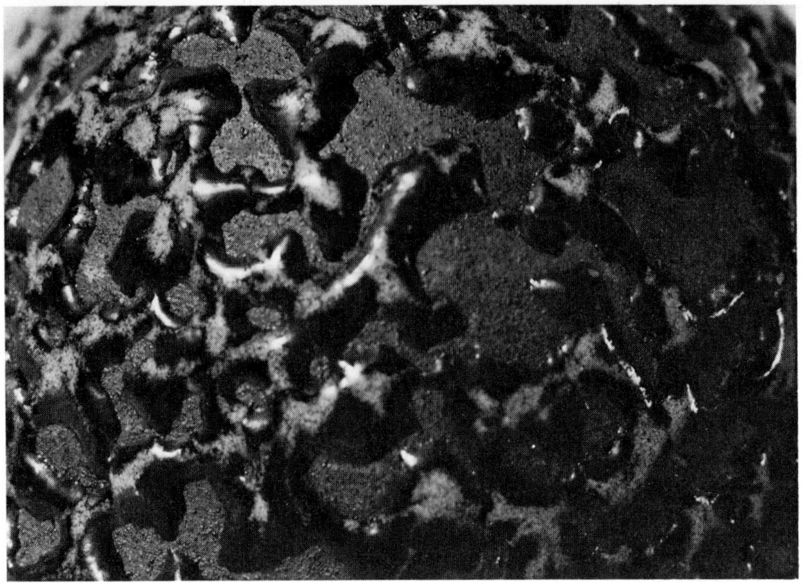

15-4 *Crawling.*

when the glaze is applied. As the glaze dries on the pot, severe shrinkage cracks appear, like mud cracks. These large cracks in the unfired glaze might initiate separation and crawling when the glaze melts.

At other times crawling occurs when the glaze recipe contains too much kaolin, or its empirical formula makes crawling inevitable. A glaze formula may be recalculated for less raw kaolin, but otherwise an adjustment is difficult. Other than care in glaze application, there is not much that we can do to avoid crawling, if the formula is faulty. Discard the recipe when crawling appears again. A pot with a crawled glaze cannot itself be cured or made better.

Matt Glaze Fires Shiny Matt glazes generally require a thicker coating to develop their characteristically luscious, buttery surfaces. When they are applied too thinly, many matt glazes melt in a more reflective, shiny surface. Overfiring is another cause of shiny, rather than dull matt textures. Nevertheless, if a pot seems promising and deserving of the extra effort, an additional coating of glaze may be applied and the piece fired once again.

Bare Patches It perplexes beginning potters when their work is taken from the kiln with some bare portions where they knew they had applied glaze. The glaze coating has apparently disappeared. The flaw develops when the bisk has been severely underfired and is overly porous or if the bisk was not washed well enough and remains dry and dusty. When the glaze was applied, its water was so rapidly absorbed into the bisk that the glaze dried in a layer separated from the pot's wall. Subsequently, during the early stages of firing large sections of glaze came crumbling away from the pottery surface, falling down into the bottom of the vessel or onto the kiln shelf (another good reason to have kiln wash on kiln shelves). A pot with this defect is difficult to retouch or reglaze successfully.

The pot is returned from the fire almost as it was when it went in. The glaze is chalky and dry, almost or completely unmelted. The kiln may have been very much underfired. However, should the cones indicate that a correct firing took place, the culprit is undoubtedly the glaze chemistry or a mistaken glaze preparation. The formula is bad, or there were errors in arithmetic and weighing. Usually, the pot cannot be salvaged. This is not often a glaze problem where commercial glazes are used.

16

TYPES OF POTTERY

The nature of things in pottery is that clays and glazes are formulated to be fired at predetermined cones. The technology requires order and consistency in the firing temperature, one that has been determined as proper for the ware. Predictably, pieces turn out best at the cone that suits them, rather than with one chosen at random. Furthermore, potters have tended to work within a favored cone range. Over long periods of time, a certain culture may favor one temperature whereas in another period or another culture a different temperature will be the norm. This practice of firing at uniform heats has made it easier for us to study pottery historically in terms of categories and types.

Countless individual pieces have been made, none of them exactly alike. They have varied in color, opacity, texture, resistance to thermal shock, hardness, durability, function, form, decoration, style, weight, and other differences ad infinitum. However, because most work has tended to cluster around just a few heat ranges, a suitable classification system is relatively simple to arrange. This consists of four categories, within which nearly all pottery can be placed: earthenware, stoneware, porcelain, and china. Virtually all firings can be traced to some heat equivalent within the range of cones 015 to 15. The four categories are determined primarily by the temperature range required to make the clay mature for maximum hardness and durability.

When one potter reveals the range of heat at which he fires his work, another potter can easily surmise a great deal: the kinds of clays probably used, the general chemical and visual characteristics of the glazes, aesthetic opportunities inherent in the heat, the hardness and durability of the product, constraints on size and form, color ranges and nuances, and limitations on function. Understanding pottery in terms of categories and generalizations is useful to potters, as well as to art historians and collectors.

EARTHENWARE

From the very beginning of pottery making until the present day, most of the work has been what we now call *earthenware.* There are two good explanations for this abundance: (1) suitable clays have been common and plentiful and (2) the heat required for these clays is readily attainable, even in early or primitive kilns. Dark-firing, native, or local clays are close to the surface and are easily taken from the earth. Many of them have been found to work satisfactorily as they are, without the complication of body alteration and formulation (see Chapter 2). These clays become reasonably hard, even when fired to a low temperature. Most earthenware has been fired to the temperature equivalents below cone 1. Relatively unsophisticated kilns do the job well and are economical at that heat level in fuel consumption or electric energy.

The natural earthenware clays all fire to some variety of buff to red-brown, the color given them by their inevitable iron content. A white-firing earthenware clay is virtually unknown. However, white earthenware bodies can be formulated around a powerful flux, such as nepheline syenite, talc, or body frit, to lower the maturing temperature sufficiently and with enough plastic substance, such as kaolin and ball clay, to make the mixture workable. The white color can be altered by adding any body colorant (see Chapters 4 and 9). Some earthenware, fired above cone 1 and reaching to cone 4, is composed of body mixtures that produce a more durable and harder ware similar to stoneware.

Earthenware is opaque. Without a glaze to cover it, most earthenware is neither waterproof nor impervious. Water in the vessel would slowly seep through the wall. In the broken cross-section the fired clay appears open grained and coarse, like a cracker. It might absorb as much as 10% of its own weight in water, is the least hard of the four types of pottery, and chips and breaks relatively easily.

Earthenware glazes are formulated to melt at the ware's characteristically low temperatures. Therefore, they are comparatively soft, scratch more readily, and decompose more under attack by acids and abrasives. At these low temperatures, glaze colorants become brighter and more intense, compared to the hue tendencies at higher heats.

The earthenware style is distinctive. There is a directness of form, glaze, and color. The heft of the ware, its feel of substance in the hands, may be more tentative, vulnerable. Its sense of being does come, appropriately, from the earth.

Historically, probably the best known earthenware is called *majolica.* Majolica is typically made of iron-bearing clays ranging from buff to red-brown in color. The characterizing glaze is very opaque and bright. Majolica is always known by decoration painted into the raw, dry glaze coat, before the glaze is fired. Upon being fired, the decorative colorants soak into the white glaze, in a fusion which is unique in its visual integration. Traditionally, a lead glaze is used, made opaque white by tin oxide.

This means of making glaze opaque was invented more than a thousand years ago and brought to Spain with the Moorish invasion by the 8th century after Christ. (The Mediterranean island of Majorca is the root for the name.) Potters of the Italian Renaissance attained the height in majolica artistry. The technique immigrated to the north of Europe, where it became known as faience in France, and Delft ware after the Dutch pottery center. Today, the term "majolica" is often applied generically to describe colorful, painted decoration applied to the unfired surface of a white, opaque glaze. It has great appeal for earthenware potters.

RAKU

Raku is a very special type of earthenware with a unique place in contemporary pottery making. It is a Japanese method of working with clay that goes back more than 400 years. It originated as an integral part of the Japanese tea ceremony. The preparation and serving of tea in special pottery vessels made in that particularly congenial spirit was an expression of the dominant Buddhist theology and its manifestation in a national life style. Symbolic raku forms, glazes, textures and colors came to be treasured as art works. Raku masters handed down the traditions of their art and craft.

In this country, accurate information about raku was only first published in the years immediately following World War II. Since then there has been a rapidly spreading and growing interest in making pots in the raku style. To some, the great popularity of raku is a passing fad, or at least a cult. The term "raku" remains in force, but the American melting pot has made any connection of contemporary American raku work with the traditional Japanese raku pottery often vague.

The process is immediate and participatory. Pots, bisked before the raku firing, are glazed and thoroughly dried. If possible, they are warmed around the kiln. The crucial element is that the kiln has been pre-heated to a good redness, somewhere in the range of cones 015 to 08. Depending on the type of kiln used, the pots are placed in the hot kiln one at a time or several at once. In a short time, perhaps 15 to 30 minutes, the glaze, viewed through the peephole, melts. The sight of molten glaze is unmistakable, surprisingly resembling thick, wet, glistening ice. At this time the pot is removed from the kiln with steel tongs.

The pot may be permitted to cool on a nearby kiln shelf, or it may be quenched in water; glaze variations result. Placed deep in a container of some combustible material, such as seaweed, sawdust, leaves, or wood shavings, the glaze develops interesting reduction effects. For some, the appeal is the action and participation, and its aesthetic value derived from the process.

As in all pottery categories, raku's aesthetic character is shaped at least to some extent by the details of the raku technology. The following conditions are suggested:

1. The clay used should be coarse grained, a sandy or heavily grogged mixture, for the piece to endure the thermal shock of rapid fire and sudden quenching and cooling.

2. The form ought to be simple, with little in the way of complex curvature and detail. A simple form is more at home with the rough clay required. Uncluttered shapes are apt to be more resistant to shattering under the stresses of the procedure.

3. The glaze must be formulated both to melt well at very low temperatures and to cool quickly without major flaws developing. However, crazing is almost certain and sought after. Because they melt so well and have such lovely visual properties, lead-silica glazes had been used from the very beginning. Some American potters now use leadless glazes. Either way, raku glazes are usually applied thickly, a dimension which obscures much detail and definition beneath the glaze. This is another reason for simplicity and directness in shape and treatment.

EQUIPMENT AND MATERIALS

1. A small kiln capable of reaching the desired heat of cone 015 to 08 in a few hours, and the capacity to retrieve the heat range lost when the kiln is opened for pots to be removed. In the Japanese tradition, small, portable wood or charcoal kilns were used. Much raku is now done in small studio electric or gas kilns. A raku kiln can be improvised from a steel drum lined with ceramic

16-1 *A traditional wood-burning raku kiln with sagger.*

16-2 *Raku firing at the Vermont State Craft Center. The kiln is a steel drum lined with ceramic insulating fiber.*

Used by permission of the Vermont State Craft Center at Frog Hollow, Middlebury, VT.

fiber blanket and fired with propane fuel. Any such structure should have an efficient method to open the kiln for removal of finished pots.

2. Some kiln shelves placed on the kiln or close by, to dry and warm the pots prior to firing and to hold red-hot pots fresh out of the kiln.

3. Steel tongs for stacking and retrieving work.

4. Heat resistant gloves or mitts.

5. Bisk pots made from a grogged or sandy body. Frequently, the potter's stoneware clay works well enough in the low-fire bisked state. Pinched, coiled, slabbed, and thrown work are all amenable to the process. Slipcast pots do not usually have sufficient internal strength to resist rapid temperature change.

6. Kiln shelves for the kiln, carefully and generously coated with kiln wash.

7. Low-fire glazes. In recent years lead as a glaze ingredient has been questioned because of its potential toxicity (see Chapter 19). However, the appeal of lead for its benefit to color and texture in glaze remains strong. Lead glazed raku should not be used for serving food. Even lead-free raku should be avoided for the glaze often contains flaws such as crazing, blisters, and pinholes. Sample recipes of lead and leadless glazes for raku are:

white lead	85	Gerstley borate	80
flint	15	nepheline syenite	20
	100		100

white lead	61	barium carbonate	6
frit	18	Gerstley borate	64
flint	21	Custer feldspar	12
	100	ball clay	6
		flint	12
			100

Some variety of binder added to the liquid glaze mixture helps the raw glaze adhere to the pot. White glue, sugar syrups, gum tragacanth, and a number of available synthetic gums and binders work well to give a crusty glaze coating. Raku glazes are easily colored by adding some colorant to the basic glaze recipe (see Chapter 9).

THE RAKU SEQUENCE

1. Pre-heat the kiln to the desired cone.

2. Apply glaze to the bisked ware. The glaze thickness should be generous. Dipping and pouring are good application methods, although brushing and spraying may also be used.

3. Rest the glazed bisk on and around the kiln, to dry the glaze and warm the pots.

4. When the cones have melted properly, the pots are set in the kiln. The traditional raku kiln has room for one piece at a time, but in other arrangements several pots can be fired simultaneously. Steel tongs are used to grip the work, and heat-protective gloves should be worn.

5. Observe the effects of heat on the glaze, looking through the kiln's peephole.

6. The glaze is thoroughly mature and melted when it turns highly glossy, when it reflects light like ice in the sun.

7. Gripping one piece at a time with tongs, remove the red-hot work from the kiln. If the tongs are taken quickly from the pot, the tong marks in the glaze heal over as the glaze cools.

8. There are alternative methods to treating the glaze. The red-hot piece may be placed on a kiln shelf and left to cool. To induce more dramatic crackle patterns, the work may be quenched immediately upon retrieval from the kiln, directly into a pail of water, buried while still glowing in some smoke-producing material, such as seaweed, leaves, straw, charcoal, sawdust or wood shavings. Reduction colors, textures, and lusters are encouraged by this procedure. The smoke-inducing ma-

16-3 *Dipping a fired pot in wet leaves.*

16-4 *Quenching a fired pot in water.*

terial should be in a metal pail or barrel and covered over quickly to encourage reduction by smothering and burning.

STONEWARE

Stoneware was made possible by the discovery of clays capable of withstanding high heat and, consequently, the development of new kiln designs and firing techniques to mature these clays. The first stoneware was probably made in China more than 2000 years ago. In the history of pottery technology it was a large step forward, towards a greater hardness than that of earthenware. There is a simple definition of stoneware: Pottery which achieves maturity somewhere within the range of cones 4 to 12.

The name "stoneware" is appropriate, for it has strong visual and tactile references to rock. It is an opaque product, in studio pottery generally colored warm grey, buff, or red-brown. The fired ware is dense and hard. Unglazed it absorbs less than 2% of its own weight in water. Nevertheless, it requires a glaze to make it entirely waterproof. Stoneware glazes are also fired to the same high temperature. This results in a glass coating that is very hard and fused with the clay surface beneath.

While stoneware clays are not as commonplace as earthenware types, many deposits are mined. They are surprisingly similar in how they can be made into solid, strong, stone-colored, rock-hard pottery wares. These clays are often satisfactory used directly as they come from the earth, without alteration. However, potters may prefer to assemble stoneware body compositions to meet individual or unique requirements, in terms of visual and tactile properties, as well as those of plasticity, porosity, and vitrifiability. Formulated bodies can be made white or any color. But most studio stonewares tend to emulate the colors of the natural clay. Stoneware is particularly amenable to the visual and textural effects of reduction firing and salt glazing. The mastery of stoneware clays, techniques, and their high firing preceded the development of porcelain in China.

PORCELAIN

Porcelain is considered by many pottery connoisseurs the supreme technical achievement. It certainly is the ware type that receives the highest fire, generally within the range of cones 10 to 15. Porcelain is always a white body and completely dense. Even without a glaze it is impervious or waterproof. Porcelain bodies are formulated to fuse or approach the melted state at that high heat; the ware gets close to the glassy condition. Therefore, when it is thin in cross-section, as it is in dinnerware, it is translucent. Light appears to filter through. Porcelain collectors also appreciate its clear, bell-like tone when struck.

There is no such thing as a single, naturally occurring porcelain clay. It is always a formulated body, consisting of a plastic kaolin, feldspar, and flint. For the purpose of maximum whiteness, the kaolin may very well be a residual variety. The combination does not result in great plasticity, consequently small additions of ball clay or bentonite may be made for improved workability, but this diminishes its essential white purity.

Porcelain glazes are composed of the same ingredients as the body but in different proportions to facilitate complete melting. The glaze is fired together with the pottery to the end cone. Body and glaze are fused as if one. True porcelain glaze is hard enough to withstand scratching, except by a diamond. When color appears in porcelain, the hue is in the glaze. It is a marvel that the standard for technical and aesthetic excellence is the porcelain of the Chinese Sung Dynasty of a thousand years ago, exemplified by celadon glazes on immaculately conceived and executed works.

Each of the pottery-ware types has a unique personality, and porcelain is no exception. Its technical constraints are substantial, a white body not easily worked and subject to threatening, extremely high heat. They challenge the sensitive potter's discipline and creativity. Some porcelain forms are spare and trim, and others are distinguished by exuberance and lavish workmanship. Nevertheless, its technical and mechanical character nurtures a distinctive sophistication and elegance in form, glaze, and decoration.

CHINA

Potters make earthenware, stoneware, and porcelain, but china is an industrial product. A description is included in this book on studio pottery because of its historical relationship to porcelain, and because it might be at least of passing interest to consumers and collectors of tablewares. China ware is a child of the Industrial Revolution, having been conceived in the mid and late eighteenth-century England to fill the need for a high quality, mass-produced ware. Essentially, china was a modification and adaptation of porcelain, the large-scale manufacture of which for a mass market had proved impractical.

Actually, china refers to a type of pottery matured in the bisk firing most often in the range of cones 6 to 12. The glaze firing is at a lower cone, somewhere between cones 1 and 6. This was one of the innovations to make factory production practical. The china product is completely or virtually waterproof, even before the glaze is applied. It may or may not be translucent, depending on body formulation and wall thickness. The glaze, fired at a lower heat than porcelain, is softer than a porcelain glaze. Consequently, it scratches more easily. China wares include household dishes and sanitary wares for the kitchen and bathroom. The word "china" has become popularized; it often is used generically to include all factory-produced pottery wares whatever their factual classification, as in "China Departments" in stores and gift shops.

17

PLASTER IN THE POTTERY STUDIO

PLASTER

Plaster is a work horse around the pottery studio. It is an ideal substance for making wedging boards and bats. Water saturated, thick, plaster slabs keep the air humid inside damp cabinets. Potters use plaster drying bowls to convert slurry into plastic clay; the porous plaster absorbs a great deal of excess water. But the chief use of plaster for pottery production is in molds for slip casting. Plaster can also be made into molds for press casting and slump casting of plastic clay.

We know that plaster was used by the Egyptians more than 4000 years ago. The ancient Romans set great store by it as a building material. Large parts of Paris are built overlying gypsum beds and old gypsum mines. Gypsum is the raw material of plaster. The term "plaster of Paris" probably originated from that location of rich gypsum deposits. Although it has been a much used art and construction material for a very long time, plaster did not come into service for potters until recent centuries. Earlier, most molds were made of soft bisk.

Plaster is a ceramic material. It is a mineral of the earth and is processed by intense heat. After the gypsum rock has been mined and crushed, it is heated in a special kiln. The heat drives some of the chemically combined water from the gypsum molecule and reduces the rock to a soft substance. The calcined material is milled into a fine white powder or plaster. The essential usefulness of plaster derives from its great thirst for water, its desire to reconstitute itself into a complete molecule. When mixed with water it sets in a short time into a hard, white solid, very much like the original gypsum rock.

Plaster is easy to shape in various stages of consistency. In the slurry or liquid state it may be poured within a frame, and upon setting it keeps the shape of the frame. For example, a bat can be made by pouring plaster into an empty pie tin. In the "cheese" or semi-solid state it can be formed handily; once hard it can be scraped, cut, and chiselled.

Plaster is a stable substance. It does not shrink or expand, or otherwise change dimensionally. It is a refined and smooth material, properties that permit molds to reproduce meticulously details, contours, and surfaces. To the potter's dismay, flaws may be reproduced as well. Not the least of its virtues, plaster is plentiful and inexpensive.

Probably the most practical property of plaster is its porosity or absorption, readily taking water into its pores from the slip or plastic clay lying next to it. The casting slip sets into a solid skin against slip-mold surfaces. Plaster in the form of a bat, slab, or wedging board dries plastic clay that is too moist. Soft pots do not stick to plaster bats, nor does clay being wedged stick to plaster wedging surfaces, all because of the porosity of the material. Many potters prepare clay by first combining all the body ingredients into a slurry, following which the slurry is dried to the plastic state in large, plaster drying bowls. This is a very good method to make clay (see Chapter 5). Plaster bats and wedging tables have a long life in the studio. Press molds and slip molds can be used repeatedly for scores of castings.

Plaster companies manufacture a variety of plasters for diverse uses. A plaster similar to No. 1 Pottery Plaster works best in the potter's studio. A supply no larger than can be consumed in a month or two should be purchased. Store the plaster in a clean, tightly closed container to protect it from atmospheric humidity and the debris of the studio.

PREPARATIONS

A suitable compromise between ideal absorption and ideal hardness results when 2 3/4 lbs. of plaster are mixed with one quart of water. For special purposes this ratio may be adjusted slightly: More water produces a more absorbent but softer plaster; less water gives a harder plaster that is less porous.

The amount of plaster to prepare is determined by the volume to be filled. One quart of water plus 2 3/4 lbs. plaster fills a volume of 80 cubic inches, no matter the configuration of the volume, cubic or spherical. To obtain the volume of a cubic form, multiply the length in inches × the width in inches × the height in inches. The calculation for a cylindrical volume is 22/7 × radius^2 × the height in inches. Divide the total by 80 to obtain quarts of water, and multiply by 2 3/4 for pounds of plaster. For example:

PLASTER-WATER RATIO

Cubic Inches	Water	Plaster
20	½ pt.	11 oz.
40	1 pt.	1 lb. 6 oz.
80	1 qt.	2 lb. 12 oz.
160	2 qt.	5 lb. 8 oz.
240	3 qt.	8 lb. 4 oz.

If the studio is spacious, there will be room enough for a separate, segregated area in which plaster work can be conducted without risk of contaminating the other materials in the clay area. But, should the studio be limited in space or plaster work done so rarely that a separate section would be extravagant or wasteful, line a work table well with newspaper. Exercise extreme care that no plaster falls or drifts into the clay or glaze. To avoid a host of firing problems, such as pitting and spit-outs, plaster must be scrupulously kept from contaminating all other materials in the studio.

Have ready a scale, such as a platform kitchen scale, one that registers in pounds and ounces. A liquid measure for water, and a large, smooth plastic bowl or pail for mixing the plaster will be needed. The frame, dam, or cottle into which the plaster mix will be poured should be readied in advance and well secured against leaks.

Immediately following the pouring, the mixing bowl or pail will have to be washed at once, before plaster residue hardens and sticks. Hardened plaster is difficult if not impossible to remove. Plaster should *never* be emptied into sink drains for fear of permanently plugging the plumbing. A good cleaning arrangement is to have a large container ready, such as a garbage can half full of water. Plaster-laden tools and mixing bowls, as well as the potter's hands, may safely and effectively be washed without danger to the plumbing. Eventually the pail becomes full of unsalvagable waste, and the whole business consigned to the dump.

COTTLES AND FRAMES

In the pottery studio, plaster is nearly always shaped while in the liquid state. Therefore, some means

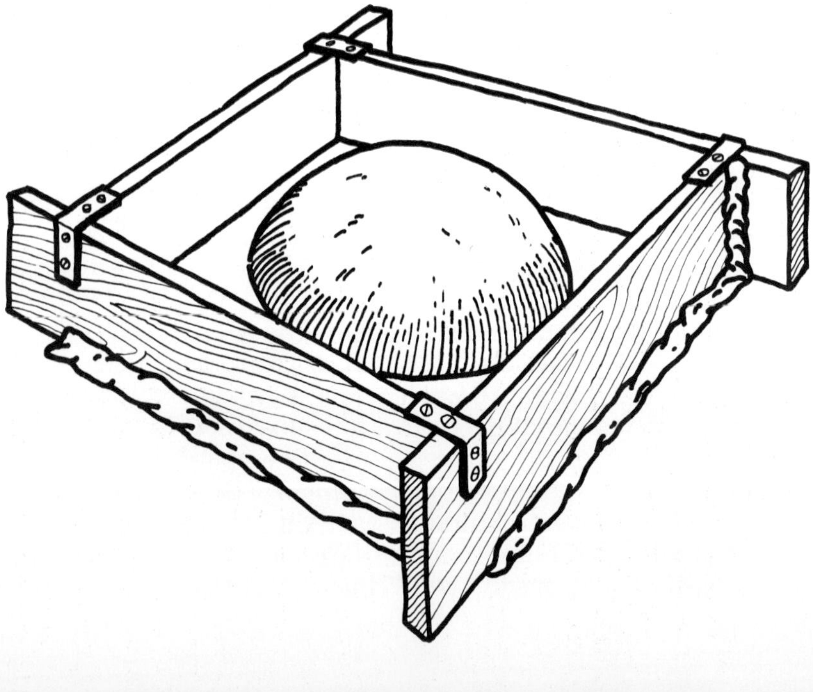

17-1 *A cottle and clay mound arranged for the pouring of a plaster drying bowl.*

of containing the plaster must be provided. A kind of dam or frame called a *cottle* is improvised, which holds the liquid plaster together until it sets hard. Pie tins or shoebox lids are simple enough forms into which to pour plaster for handbuilding bats. Wedging boards, large slabs and shelves, press and slip molds require cottles that are more substantial and sophisticated in their construction. Wooden boards are selected that are wide enough and long enough to fully frame the volume on all four sides. They are interlocked at each of the four corners by using brackets or nails. Where plaster slurry might conceivably leak through, coils of clay are pressed along the seams and cracks, at corners and along the work table. Round cottles or dams are easily made from sheet-metal strips curved to the desired circumference and bound tightly with cord. Small cylindrical cottles can be made of cardboard. Clay coils seal the seams where the cottle joins the table and wherever plaster might seep through.

Plaster normally adheres to the surface next to it but more so to porous surfaces, such as cardboard or unpainted wood. It releases much more readily from closed surfaces, such as metal, plastic, and sealed wood. Therefore, cottle materials and table tops require preparation to prevent sticking to plaster. Wood should be given a coat of shellac, followed by a light application of some oil. Metal cottle materials require merely a thin slick of oil for easy release from set plaster. Motor oil, lubricating oil, and cooking oil work equally well.

MIXING AND POURING PLASTER

In a clean mixing bowl or pail put the desired amount of water. Use clean water from the tap, preferably at room temperature. Warm or hot water should not be used because it accelerates setting time too rapidly. Weigh the desired amount of plaster in a separate container, having adjusted the scale for the weight of the container.

Sift or slowly scatter by hand the plaster powder evenly over the water. Avoid dropping handfuls all at once into the water, as large lumps would be created. The sifted and scattered plaster is quickly taken up by the water.

When in a few moments all the plaster powder has disappeared into the water, it may be considered wet through or *slaked*. Allow the mix to sit undisturbed for a minute or two, important for having a smoothly pouring, plaster slurry. Now stir the mixture by hand, steadily and around from the bottom to the top. Stir slowly, not to beat or whip air bubbles or froth into the slurry. Proper mixing tends to carry air pockets upwards to the surface where most of them break apart. Bubbles remaining at the surface break when the potter gently blows at them.

Continue stirring until a smooth, uniform, creamy mass is produced. Shortly, the plaster slurry appears to thicken ever so slightly. This is the climactic moment, the right time for pouring into the cottle. Do not hesitate; it may set in midstream. Pour slowly and steadily, without splashing, to avoid creating air bubbles within and entrapping air pockets. Continue pouring in a continuous stream until the volume is filled to the designated level. Immediately wash the mixing vessels, tools, and hands in the waste pail.

Potter's plaster can be expected to set solid approximately 25 minutes after the powder and water have been combined. Setting time can be abbreviated with more vigorous stirring. Conversely, more leisurely stirring increases the time before hardening. In some mold-making procedures plaster is

17-2 *Slowly sift the plaster powder over the water.*

17-3 *Stir the mixture slowly.*

17-4 *Pour the plaster steadily, without splashing.*

worked in the "cheese" state. This condition between liquid and solid may be extended a few minutes longer if there has been a minimum of hand stirring. Setting plaster is accompanied by the generation of heat, and the solid plaster feels warm to the touch. When the hard plaster has cooled, it is considered set.

Remove the cottle material from the solid plaster form. Sharp edges and corners should be trimmed or chamfered, using a potter's knife or rigid scraper. The purpose of the treated corner is to minimize chipping.

The mixing process has incorporated a large quantity of water in with the plaster. For most studio uses, such as bats, slip-casting molds, drying bowls and slabs, and press molds, the water is encouraged to evaporate from the set plaster. Arrange the piece for maximum air circulation under and around its surfaces. While warm air is an effective drying agent, care should be taken not to heat the plaster above 110°, for it tends to deteriorate above that temperature.

BATS, WEDGING BOARDS, AND RELATED MATERIALS

Small plaster bats 6″ to 9″ in diameter arc so useful around the pottery studio, it sometimes seems the potter never has enough of them. The most convenient way to make these bats is to pour the plaster into lightly oiled pie tins, usually leftovers of frozen foods. Square bats can almost as easily be made in a frame of 1″ × 3″ × 12″ cottle boards, held together at the corners by a nail.

Large plaster slabs for drying clay or for damp cabinet shelves require cottle boards larger than

17-5 *A sheet-metal cottle for a drying bowl.*

17-6 *Preparation for making a plaster drying bowl.*

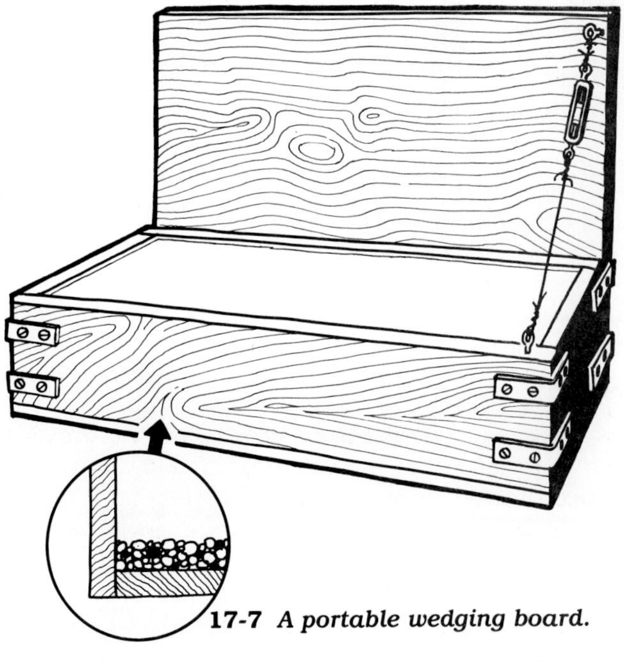

17-7 *A portable wedging board.*

the slab's dimensions. The boards interlock at the corners by means of metal 'L' brackets to hold the cottle intact against the force of the liquid plaster. The pressure from within is substantial and should not be underestimated. These cottle boards may be used repeatedly and adjusted to make slabs of varying dimension.

If the plaster slab is quite large, some thought should be given to its reinforcement. When half the thickness has been poured, lay down some burlap, chicken wire or screening, then complete the pouring to cover the reinforcing material.

Plaster bowls for drying clay are most effective when the plaster substance is at least 2½″ to 3″ thick. These bowls are ordinarily used to convert clay slurry into a usable plastic condition. Therefore, their extractive powers must be prodigious.

You might be tempted to make very large bowls, but they might prove too heavy to lift. Choose a diameter between 16″ and 20″. First make a mound as the mold for the concave portion of the drying bowl. The most convenient material to use is plastic clay; it is easy to shape and will not stick to the plaster. Care should be taken to prevent the moist mound from drying beyond the leather-hard stage. Bend sheet-metal strips in a circumference around the mound, leaving about three inches of space in between for plaster thickness. Bind the metal cottle securely, with stout cord. Seal the seam at the bottom edge all around with generously applied coils of clay. The force of plaster to leak through is substantial. In the absence of sheet-metal cottle material, wooden boards that interlock or fasten at the corners may be substituted and will work just as well.

There are no standard or uniform dimensions for wedging boards. The wedging surface should be large enough in area for the potter to thump, pound, smack, and knead large masses of clay. The working surface should be situated about 32″ from the floor, a height to facilitate good work and efficient leverage. However, adjustments can be made to suit the height of the potter. The plaster slab ought to be at least 4″ thick but preferably closer to 6″. The frame in which the plaster is poured is a permanent part of the wedging board structure. Therefore, secure the boards at the corners with screws, bolts, and corner braces for enduring strength. An inch of gravel or crushed rock may be placed as a base for the plaster. It may be difficult to mix and pour a large quantity of plaster all at one time. Pour in two batches, and deeply puncture and scratch the first layer before it sets, to act as a binding grip on the second layer.

After the plaster has set, attach and securely brace a vertical arm to the rear frame, about 18″ tall. Attach a cutting wire, made of .003″ music wire. The wire is held in place at both ends by means of strong hook eyes, and is made taut by a common hardware turnbuckle.

Since wedging boards are usually extremely heavy and difficult to move around, it is most practical to construct this piece of equipment at the location desired for wedging. Some potters might prefer to have a wedging table with its own legs attached. Use heavy construction-grade lumber for the legs, substantially braced for a good, solid structure.

18 FURNISHING THE POTTERY STUDIO

The tools and materials essential for working with clay are few. A plastic clay or body, a good glaze or two, a few hand tools, and a suitable kiln constitute the absolute necessities. However, the creative lives of potters are almost always more complicated than the spare list above suggests. The more a potter gets to doing things self sufficiently, such as formulating and preparing clays and glazes and extending the range of creative activity, the more the potter seems to be drawn to the nub of personal involvement. But at the same time there is the anomaly of soaring growth of the studio inventory.

A fledgling potter would be greatly assisted in setting up a studio if there were lists of materials and equipment and if the items were arranged logically according to work in the pottery studio. Such lists are given below. From them the beginning potter may pick and choose, for beginning simplicity to greater involvement.

Materials are referred to by their generic names. Pottery minerals are so variable in their natural form that at least slight differences may be expected within the same category and in the frame of time. For example, there are several ball clays available. Should your recipes specify an individual ball clay, it is best to stock that one. If it becomes necessary or convenient to substitute, prudence indicates that the recipe be tested with the substitute before making large investments in time, energy, and material. Some body or glaze recipes merely indicate the generic name for an ingredient, such as "kaolin", "ball clay", or "feldspar." Use the brand of the material stocked in the studio.

Pottery-supply catalogs are excellent sources to learn about availability and costs. Regional and national supply companies regularly advertise in *Ceramics Monthly*, *American Craft*, and other journals. They often furnish a prospective buyer with a catalog at little or no cost.

CLAYS

The decision as to which clay to purchase is based on more than one consideration. There are questions of color, texture, suitability for specific uses, how the pots will be made, and whether one clay will serve the general purposes of the studio or there will be a need for more than one. Probably the central factor in the choice of a clay is the firing range of your kiln. A kiln with an upper limit of 1950° determines that the work of the studio will be earthenware. The most frequently used firing ranges cluster around cones 06 to 04 (earthenware), cones 4 to 5 (low stoneware), and cones 8 to 10 (stoneware and soft porcelain). The potter's next decision is whether to compound the body in the studio or to purchase the material completely tested and prepared ready for use. Some natural clays sold are quite satisfactory for work as they are, while others may need adjustments of one kind or another. Many beginning potters would rather spend their time and energy making pots than preparing clays. Clays must also be stocked for use in engobe, slip, and glaze recipes. Some that should be considered are:

Albany clay For high-fire slip glazes.

Ball clay In body formulations to increase plasticity; sometimes present in glaze recipes.

Bentonite Very small percentages for plasticity in body and slip formulations; in glazes to suspend coarse ingredients in the watery mixture.

Fire Clay Valuable component of high-fire stoneware bodies for properties of strength, workability, texture, and fired color.

Kaolin Either a secondary or primary kaolin, as an ingredient of nearly all glaze recipes and most white clay bodies. Combined with an equal weight of flint to make kiln wash.

Red clay Low-fire secondary clays, many of which are satisfactory alone or useful as a body component for contributions of plasticity, iron color, and fluxing.

Stoneware clay Goldart, Monmouth, and Jordan are the names of some popular stoneware clays that work well as they are. Potters also use them as the dominant clay in a formulated stoneware body.

CHEMICALS

Antimony oxide Yellow colorant and opacifier in glazes.

Barium carbonate Flux in high fire glazes but good for matt textures in low-fire glazes.

Cornwall stone Feldspathic flux for high-fire glazes.

Cryolite A sodium-fluoride flux for special glaze purposes.

Dolomite A calcium-magnesium flux often appearing in medium- and high-fire glaze recipes for characteristic magnesium matts and good influences on color.

Feldspar Either as a soda or potash feldspar, it is a mineralogical composition of alkalis, alumina, and silica. A studio necessity, the potash feldspars appear in most glaze recipes and the soda types in many clay-body recipes. They are known by their brand names, such as F-4, Oxford, and Custer.

Flint Powdered quartz, the major source for silica in nearly all glaze and body recipes: combined with kaolin to make kiln wash; a good bedding material for kiln stacking; a useful parting agent for slip casting; a studio necessity.

Gerstley borate The major source for insoluble boric oxide in low-fire and medium-fire glazes. Colemanite was formerly used, but now Gerstley borate has taken its place.

Lead monosilicate A frit of lead with silica to flux low-fire glazes. As a source of lead, it is safer for the potter to handle than raw lead chemicals.

Magnesium carbonate The major source for magnesium oxide in high-fire stoneware glazes. A good influence on color, especially cobalt.

Nepheline syenite A feldspathic type of flux for low-fire bodies and intermediate-fire glazes.

Spodumene A good source for lithium in some glazes and flameware bodies.

Talc A powerful flux for low-fire glazes and bodies.

Tin oxide The traditional opacifier for glazes. Probably the preferred opacifier for its whiteness and effect on color but much more expensive.

Whiting The major source of calcium oxide for glazes.

Zircopax and Superpax Zirconium compound opacifiers for glazes. Less potent than tin oxide but also less costly.

COLORANTS

Chromium oxide Greens.

Cobalt oxide Blues.

Cobalt carbonate Same colors as the oxide but less potent. It mixes more easily.

Copper oxide. Greens. Source for copper reds in certain reduction glazes.

Copper carbonate Same colors as the oxide but less potent. It mixes more easily.

Ilmenite, powdered Fine-grained, dark speckles.

Ilmenite, granular Coarse-grained, large, dark speckles.

Iron chromate Greys.

Iron oxide, red Straw to red-brown colors.

Manganese dioxide Characteristic browns to lavenders.

Manganese carbonate Same colors as the oxide but less potent. It mixes more easily.

Nickel oxide Greys, grey-greens.

Nickel carbonate Same colors as the oxide but less potent. It mixes more easily.

Rutile, powdered All together, it opacifies, gives a buff color and a mottled or broken matt texture. The powdered rutile is preferred. A versatile material for use in glazes and bodies.

Rutile, granular Dark speckles with some of the influence of the powdered form.

Underglaze For application on bisk and bone-dry clay. It comes in a wide range of color qualities in the form of powder, pencil, crayon, paint, magic marker.

Overglazes For application on fired glaze surfaces. Comes in a wide range of color qualities in the form of powder, pencil, crayon, and paint.

FRITS

Frits are pre-melted compounds of a wide variety of formulations used by potters to introduce otherwise soluble or potentially toxic ingredients into glaze and body recipes more safely and conveniently. Frits are not a staple in the cupboard of the beginning potter but may be purchased as called for in a favorite recipe; use only the frit indicated in the recipe, such as "Ferro Frit 3304."

KILN MATERIALS

Shelf post Used generally in threes to support kiln shelves. They come in 1″ gradations, 1″ to 12″, in round, square, fluted-square, and triangular cross-sections.

Kiln shelves Fire-clay shelves (cordierite) are tan colored, satisfactory to cone 8, and should be stocked if the highest firings are usually not much above cone 5. Otherwise, purchase silicon carbide shelves (black in color), but they are much more expensive. Both types come in a wide range of shapes and sizes for the studio's electric, gas, or wood kiln.

Kiln wash May be purchased ready made or mixed in the studio. A good, general purpose wash is one half flint plus one half kaolin with enough water to make a paintable slurry.

Mending cement Known also as *bond cement.* A preparation to repair and patch small cracks and chips in kiln walls. Available in powder form or prepared moist in tight cans.

Stilts, spurs, and pins Objects made of a clay body similar to that of china, to support glazed wares in the kiln, where glaze may run down over the foot ring onto the kiln shelf. Stilts come in sizes #1 to #12 and also with wire points; pins come by length.

Pyrometric cones Smaller quantities available in bubble packs and also in boxes of 50. Choose the small or standard size, depending on the size of your kiln.

Indicating pyrometer If the studio kiln does not have one attached, the purchase of a portable indicating pyrometer is suggested. It is a valuable instrument to assist in firing bisk and glaze kilns.

Heat-resistant gloves or mittens To handle wares and kiln furniture too warm for bare hands. A necessity for raku work. Non-asbestos gloves are now available and preferred.

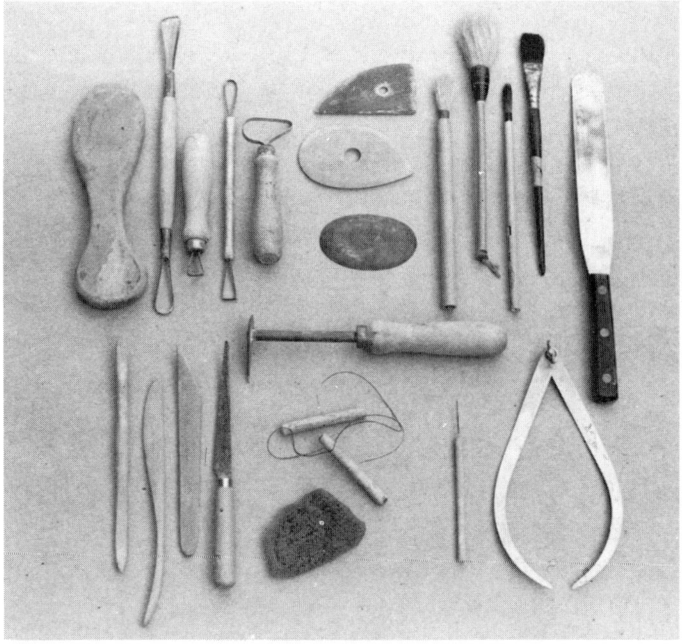

Tools for the studio.

TOOLS

Brushes
Flat-wire trimming tools
Cut off needle
Cut off wire
Elephant ear or silk sponges
Spatula
.003 music wire
Calipers
Cleanup sponges
Canvas or burlap
Electric grinding wheel
Bench whirler or Lazy Susan
Plastic bags and sheets
Plastic mixing bowls and tubs
Platform scale
Metric glaze scale
Mortar and pestle
Laboratory or pottery sieves
Rolling pin
$\frac{1}{4}''$ × $\frac{1}{2}''$ flat sticks
Throwing ribs
Wooden modelling tools
Plaster and wood bats
Small electric hot plate
Jars for storing glazes
Cottle boards
Containers for storing dry, glaze and body materials
Steel tongs

MISCELLANEOUS

Grog It is good to have some of each: fine, medium, and coarse.

Wax resist Potters make their own by melting one-half beeswax and one-half paraffin together in a double boiler. A latex type of wax-resist solution is available ready to use, not requiring melting.

Sodium silicate A deflocculant for casting slip; no other use.

Pottery plaster

Shellac and alcohol To prepare cottle boards for plaster work.

Lubricating oil

Twine

EQUIPMENT

Potter's wheel A kick wheel or electric wheel.

Kilns Electric or combustion, top loading or side loading.

Wedging board.

Damp cabinet or other damp storage Optional furniture to maintain moisture in damp clay.

Drying shelves and other shelves and racks to store work at various stages of progress.

Spray booth and compressor An optional and sizable investment, for it should be a complete system of a compressed air supply and controls, spray gun, spray booth, and exhaust system at the booth to remove waste air.

19 HEALTH AND SAFETY IN THE POTTERY STUDIO

The pottery studio ought to be a benign environment for all the creative work carried out within its walls, a safe and pleasant workplace for the potter who properly understands the materials and techniques and is prudent in their management. As with other occupations, potters should appreciate the potential physical hazards inherent in some of the materials and processes that might enter the system through inhalation and swallowing and by contact with the skin.

LEAD

It is probably the ingredient with the most notorious reputation. Lead has been valued highly, primarily as a flux in glazes fired below cone 4; it makes a significant contribution to the melting characteristics of these low-fire glaze compositions. For more than a thousand years, lead has been used in glazes for its beneficial melting qualities, its salutary effects on color, and the brilliant luster it imparts to a glaze. Unfortunately, when it gets into the human system, it is not easily eliminated. It accumulates in the body, and under these circumstances it is harmful to health.

Lead is available to the potter in many forms, but the ones most often used today are white lead (lead carbonate), the raw compound, and several lead frit formulations. White lead or any other raw lead material is the most hazardous to the potter. Whenever possible, a lead frit should be substituted because in the frit form less of the toxic material is retained in the body. Handled with great caution, lead frits offer greater safety to the potter in the preparation of glazes. The potter should be careful to keep dust down while preparing and mixing glazes. Lead-bearing glazes should not be stirred by hand, particularly if the skin is broken by a cut, abrasion, or blister. Good studio practice forbids eating, drinking, or smoking while handling these materials.

During firing and cooling, glazes containing lead give off lead vapors. Kilns should, preferably, be located in separate or secluded kiln areas, vented to the outside. If this is not possible, firing areas should be well ventilated. At the least, upon opening a newly fired kiln, potters should resist the nearly irresistible temptation to lean into the kiln, examining the results and breathing the kiln's fumes deeply.

Lead glazes may also be a health hazard to consumers who have purchased lead-glazed pieces. The lead may be dissolved by certain foods served in the vessels and then ingested. Lead is particularly vulnerable to leaching by highly acidic foods, such as fruit juices. Lead glazes may be formulated for safety in food service, by achieving a technical adjustment in the empirical formula. "Safe" glazes permit only acceptable minimal levels of lead solubility or leaching. If the potter wishes to use lead glazes for their aesthetic contributions to texture and color, only "safe" glazes should be used. In all events, these studio practices should be faithfully and routinely observed, to substantially diminish or eliminate risks to health:

1) Only use low solubility or safe frits, rather than raw lead materials.
2) Do not mix or blend lead glazes with other glazes.
3) Apply lead glazes thinly.
4) Copper upsets the stability of lead release, even in glazes with acceptable leaching factors. Do not use copper as a colorant in lead glazes.
5) Lead-glazed pottery should be fired precisely to the proper cone, neither underfired nor overfired.

Purchased or commercially prepared glazes may bear the assurance either that they are safe as far as lead release is concerned or that they are completely lead free. Any acceptably low release lead glaze, prepared in the studio or purchased ready made, may be used with confidence, providing the above precautions have been meticulously observed. However, lead-free glazes do not pose any of the risks associated with lead. All glazes designated for firing above cone 4 do not contain lead in the fired state.

OTHER POTENTIALLY TOXIC MATERIALS

A few other glaze chemicals may pose a health hazard if ingested, although it is difficult to conceive of a potter deliberately eating any of them. Nevertheless, prudent studio practice avoids eating, drinking, or smoking while preparing glazes or applying them to pots. It is far safer to stir glaze mixtures with a paddle or a rubber-gloved hand, rather than with bare hands. Barium carbonate, lithium carbonate, cadmium and selenium compounds, cobalt, and manganese are the materials that, after lead, should be approached with caution.

SPRAYING GLAZE

Spraying glaze generates a wide distribution of minute droplets over the studio atmosphere. It is actually a silica-laden mist, obviously not healthy to breathe. Therefore, glaze spraying is best done only into a spray booth with a powerful exhaust fan. The fan is designed to draw all the spray mist away from the studio, vented to a safe exit. It is prudent to wear a mask while spraying.

FIRING KILNS

The firing process normally releases carbon monoxide and sulphur compounds into the studio atmosphere. Other gases, such as fluorine and chlorine gases, may also be released during firing, as when plastic armatures within pottery are being fired or in salt-glazing. Electric and gas kilns are ideally located in separate kiln rooms, equipped with proper vents and exhausts to the outside. All combustion kilns must be vented, and many of them require connection to flues and chimneys. Electric kilns are also best served by vented hoods fitted over them, and windows opened for good ventilation during firing.

Over-eager potters may peer too closely when kilns are being fired, getting so close to peepholes as to scorch and singe eyes, eyelashes, and eyebrows. Maintain a safe distance. It may help to wear sunglasses when looking for the cones through the peephole.

HOUSEKEEPING

The particle sizes of pottery are so minute that much dust infiltrates the air and seems to float there forever. The danger to the potter breathing this fine dust may be one of several pulmonary diseases. However, the risk is of long exposure, measured in terms of many years. Good housekeeping habits should minimize or entirely eliminate the risk.

It is not difficult to keep the studio's atmosphere clean and minimize the dust. In the handling of dry materials care should be taken not to let powders fly around recklessly. Whenever possible, clean up immediately afterwards with sponge and water, and mop up as well. Sweep regularly and often but at the end of the working day, when dust raised by the broom will have an opportunity to settle down during the period the studio is unoccupied. A wet mopping is even better than the broom.

Good ventilation, such as from open windows and exhaust fans, is extremely helpful to maintain healthy conditions. Some potters have taken to wearing dust masks and respirators, but these are largely ineffectual unless they are able to filter the smallest particles of pottery dust. To be really effective, face masks should meet National Institute for Occupational Safety and Health (NIOSH) standards for filtering appropriate particle sizes.

20

AN OUTLINE OF HISTORY

As far as we know today, pottery has been made for at least 12,000 years, a chronology that ever changes as new archeological discoveries are made. In its essentials, it is little different from earlier days. As in the past, today's potters work to fulfill a wide range of human needs: mundane utility, symbolic art, self expression, spiritual representations, therapies, and the paraphernalia of social customs.

Nature continues to furnish the potter a seemingly unending and varied supply of good clays, and the major forming methods have apparently all been invented. If new technology has given us more efficient kiln designs and building materials, the convenience of electric firing, and automatic electronic controls, firing still remains fundamentally the same lengthy process that puts the potter's art to the test, and newly fired kilns are still opened as they were in past ages, with bated breath. It is noteworthy that some contemporary potters deliberately seek out the materials and methods of long ago: crude or primitive firing methods, raku, impure clays and glaze materials, handbuilding, and other inexact and unpredictable elements. If one lesson of ceramic history is that it has all been done before, then another follows: The work of the potter also reflects his time and space in never-ending diversity.

JAPAN

A few years ago I read a newspaper account of the discovery by archeologists of pencil-shaped rods made of clay hardened in an open fire. The clay sticks dated back more than 30,000 years. We shall not call that true pottery, but it is a preliminary stage. The evidence we now have of the oldest formed and properly fired pottery is from Japan. These early vessels and figurines characteristically have impressed in them a cord-like pattern or texture, from twisted rope or cord, called Jōmon pottery. The work was built by hand of coarse textured clays and fired at low temperatures. Earthenware continued to be the sole pottery type made in Japan and elsewhere for 10,000 years. The potter's wheel first appeared in Japan about 2,000 years ago, but glaze was as yet unknown there. By the 5th century after Christ the Japanese had learned from the Chinese how to build kilns capable of firing to high temperatures, and the production of stoneware commenced. These *anagama* or *climbing* kilns are still used today in Japan, and one can find a few built by modern American potters in the United States. By the 7th century after Christ glazes had come into wide use in Japan.

By these early influences—high-fire kilns and glazes—Chinese potters had already declared their dominance in the Far East. Additionally, the introduction of Buddhism (6th century after Christ) and the ritual of tea drinking (12th century after Christ), affected the direction Japanese potters would take. Some pottery-making centers specialized in preparing wares for the elaborate tea ceremony. Clays, glazes, shapes, and decorations were selected for the qualities they were perceived to contribute to the ritual. Probably, the best known of these are 16th- and 17th-century Seto, Oribe, and Shino wares.

20-1 *Japanese: early Muromachi period (first half of the 15th c.) Tamba province, Tamba kilns. A large storage jar.*

Courtesy of the Freer Gallery of Art, Smithsonian Institution, Washington, D.C.

20-2 ***Japanese: Yamashiro, Momoyana period (end of the 16th c.) Raku, attributed to Chōjiro.***

Courtesy of the Freer Gallery of Art, Smithsonian Institution, Washington, D.C.

But the name most familiar to us today is raku. Raku was originally an earthenware first made in 16th-century Kyoto, specifically for tea bowls. Kyoto and later Edo (Tokyo) developed as major centers of individual and mass production of pottery for the tea ceremony, as well as other important observances and common utilitarian wares. In 17th-century Kyoto, the potter Kenzan's work came to be so highly honored that to this day the name 'Kenzan' has been bestowed upon potters of the highest distinction. Japanese culture values a few living potters and other artists so highly as to give them the title "National Treasure."

The influence of Buddhism and the Zen tea ceremony on Japanese pottery has been the building of a tradition in which the subtleties of nature, the individuality of the potter, and the vagaries of material and technique are held to be inseparable. The most noted recent embodiment of this ideal is the late Shoji Hamada. In 1920, young Hamada had come to England with his friend Bernard Leach. There the two built a kiln and started a workshop tradition in pottery. Hamada returned to his homeland, where through his art and his teaching he spearheaded the renewal of a national folk art and craft movement. Although early Japanese pottery had been much influenced by developments first in China and later in Korea, its own tradition, now strengthened and vital, has been a source of inspiration for many Western potters.

CHINA

Archeologists tell us that Japanese pottery is older. But in technical invention and artistic influence the work of the Chinese dominated and, as many believe, ascended to the peak of human achievement. Indigenous pottery styles developed within the cultural frameworks in other countries of the Orient, but the Chinese genius furnished the technical and aesthetic standards for all, particularly during the period of some 15 centuries culminating in the Sung Dynasty, A.D. 960 to 1279. For many potters today, the work of the Chinese potters of 1000 years ago, first the T'ang Dynasty and then the Sung, is sufficient evidence that oneness of material, process, and human spirit in art is attainable.

By 2000 B.C. the potter's wheel was in use, and there is evidence of glazed pottery by 1000 B.C. A type of stoneware was being made. Firing pottery to stoneware temperatures required that these early potters had developed kilns capable of achieving at least 2200° temperature. Startling technical invention continued, with completely dense, glazed stoneware that approached the characteristics of porcelain. Reduction-fired celadon glazes were beginning to be used.

The T'ang period (A.D. 618 to 906) was in many ways a period of extraordinary cultural achievement. China had come militarily and politically to dominate much of Asia. Export and import trade flourished. Consequently, foreign influences affecting the content and style appear in T'ang art. Its pottery is a delight. By late T'ang, the great invention of that era occurs, true porcelain. Ample deposits of feldspar and pure kaolin, coupled with sophisticated control of extremely high-firing kilns, made possible the production of these completely vitreous, thin, resonant, and translucent wares. One thousand years of a culture that nurtured the art of the potter, sophistication in form and surface, technical genius in manipulation and control of materials, shaping methods, glazes, and fire, came to fruition in the Sung Dynasty.

With patronage from the court and nobility, support and appreciation from merchants, scholars, and other wealthy classes, pottery was among the many arts to flourish during the Sung period. Large, concentrated pottery centers were located in various regions of the vast Chinese empire, rather than potters scattered here and there, working separately. In these centers production for home consumption and export was profuse, while at the same time newer glazes, decorations, and forms were evolved. Instead of a monolithic uniformity throughout the nation, there were a number of regional styles. We know many of them by the names of the districts of their origin.

Sung Tz'u-chou wares are marvelous expressions in stoneware, lavish in form and rich ornamentation. The body was generally light in color, and the glaze clear or colored but always somewhat translucent. A highly stylized sgraffito decoration may be seen through the transparent glazes, for the Tz'u-chou potters first painted on the pottery with white slip, which they then flawlessly incised and carved through with line and shape. For many contemporary tastes, these strong stoneware forms and direct decorative statements have an appeal that remains fresh and relevant.

In Chien, a district in northern Fukien Province, stoneware potters in the 11th and 12th centuries turned their attention to the perfection of luscious, thick, iron-rich glazes. These glazes are much admired for their deep brown and dark plum colors, their viscosity, and their broken surfaces. In some the effect is that of thin, hair-like streaks, known as *hare's fur*. In others, the brown-black glaze broke out into silvery spots, called *oil spots*. These glazes were later adopted by the Japanese, particularly for the tea ceremony, where they became known as *tenmoku* ("ten" for "Chien") glazes. Today potters emulate these glazes for their rich surface qualities.

20-3 ***Chinese: Sung, Southern, 12th c. Celadon, Lung-ch'uan yao. Vase with two handles of fish form.***

Courtesy of the Freer Gallery of Art, Smithsonian Institution, Washington, D.C.

Chun wares of the Honan Province, with their

20-4 *Chinese: Sung, Northern, 11th c. Tz'u Chou ware, sgraffito type.*

Courtesy of the Freer Gallery of Art, Smithsonian Institution, Washington, D.C.

characteristic opalescent glazes, are also highly valued. They were first made during the Sung period, and continued to be produced for several hundred years thereafter. The hues are light blues to lavenders, often relieved by quick splashes of brighter purples and reds, but always with an intriguing pearl-like depth and reflection.

These and other stonewares reached the heights of expression, but perhaps the most sublime Sung pottery was porcelain—thin, translucent of body, white, and vitreous. Elegance of form, sophistication of surface, and celadon glaze are the hallmarks. Sung porcelain has inspired philosophical and poetic speculations about the ideals and conventions of beauty and concepts of reality.

Celadon is usually a soft grey-green to pale blue color, the result of reduction firing of glazes containing small quantities of iron. For this and for other porcelains, the Chinese had developed their kilns as precise firing instruments, mammoth arrangements of interconnecting chambers. But not all Sung porcelains were celadon-glazed. One type (Ting) was covered with a glaze ivory-like in its off-white color and satin texture. But the celadons seem to be the most celebrated, for their jade-like quality, soft texture, and color.

In time, the dominant qualities of excellence in craftsmanship and sensitivity in design came to be diluted in favor of more provincial styles. Porcelain was also transformed; the sublime purity of Sung style was lost as interest turned to sheer virtuosity in glaze, decoration, and manufacturing methods. In later periods ever more refined elaboration defines the aesthetic.

From earlier times meticulous control over copper red hues had been developed. From the Persians (Iran), the Chinese learned how to decorate with cobalt blue on white glaze and themselves mastered the use of multi-hued glazes. Their innovation was to apply colored lead glazes to fired porcelains, which were then fired a second time but at a lower temperature for the lead glazes to melt. These low-fire, opaque, lead glazes are called *enamels.* The enamel work in the Ming (A.D. 1368–1644) and K'ang-hsi (A.D. 1662–1722) periods is particularly outstanding.

Trade with the Western World had begun in the 16th century. Chinese porcelain, colors, and decoration were soon to become among the most prized possessions of the wealthy classes in Europe. Yet it was not until the early 18th century that European potters themselves learned how to make true porcelain.

THE NEAR EAST, GREECE, AND ISLAM

It is not known exactly when properly fired pottery was first made in this part of the world. Some findings have been dated back at least 7000 years. Certainly by 4000 B.C. in various regions of Mesapotamia, vertical kiln structures were in use and appeared soon after that in Egypt. Perhaps primitive by contemporary Western standards, these kilns nevertheless permitted potters to achieve a harder and more durable fired ware and to arrange for color variations in clay surfaces. Among the very first articles fired were symbolic figurines—goddesses, and fetishes—as well as utilitarian wares: urns, pitchers, bowls, ewers, tile, and the like. Command of the potter's wheel was widespread by 3000 B.C.

Much of this early work was decorated by incising, stamping, and painting with slips. It is thought that the slip painting was employed at least in part because it left an otherwise coarse surface smoother and made the vessels resistant to water. New archeological discoveries inform us that lead glazes were already known to the Mesapotamians by about 1500 B.C. Somewhat earlier, Egyptians had invented what we now call "Egyptian paste," in which an alkaline flux was mixed with a sandy clay, and upon being fired a glassy coating percolated to the surface. They had also begun some work with alkaline glazes but met with severe technical challenges. The discovery of how superbly lead enhanced the behavior of low-fire glazes was, therefore, a leap forward in the potter's power to create wares of greater beauty and improved utility.

The earliest Mediterranean civilizations of Crete, Mycenae, and other areas of the Aegean were profuse producers of pottery. They used the potter's wheel with great skill. Free-standing, updraft kilns had been improved in efficiency. These types of kilns are still seen in the region. For some reasons, although lead glazes were known in Mesapotamia about this same time, these Mediterranean potters practiced slip painting but did not, in this early period, glaze their work. The slips at first were quite ordinary mixtures of clays and water, but soon they had learned to distill only the finest sediment from the liquid clay. The refined coating made their earthenware less porous and also enhanced opportunities for more sensitive drawing and painting on the clay surfaces.

By the late Bronze Age, about 1500 to 1100 B.C., the potter's art flourished on the island of Crete. We know the subject matter of their decorations, especially nature in plant and marine life, but enjoy most the exuberance of their expression. By about 1200 B.C. Mycenae in southern mainland Greece came to dominate the region; the traditional pottery shapes and decorative motifs followed along, but animal and human figures made their appearance in the subject matter. Much pottery from this period of active trade has been found all over the Mediterranean region, both buried underground and in the sea.

By the 8th century B.C., the mantle of high civilization had passed to Greece. Greek pottery also pointed the way in the innovation of new forms and surface treatment. We are familiar with their characteristic vessel forms and story-telling decorations. The innovative shapes had originated for utilitarian purposes: to carry water, mix wine, store grains, hold oil, and keep perfumes intact, as well as to use for eating and drinking.

Greek artists painted on the pottery surfaces with increased virtuosity. They painted with ultra-refined slips—the same slip coatings as were later used on the *terra sigillata* of the Roman era—and incised through the slips. Mythological subjects

20-5 ***Greek: black-figure, 6th c.*** B.C. **Kylix (drinking cup).**
Courtesy of the Cooper-Hewitt Museum (Smithsonian Institution), New York, NY.

were particularly popular, although other human and animal forms were also depicted.

The Greeks developed the technique of low-fire reduction to a high craft, creating impressive black-red color contrasts on their pottery surfaces. This pottery was sophisticated in design and composition, its surface decoration being arranged in such a way as best to complement its form. By the 4th century B.C., Greek figured pottery had significantly declined in quality and quantity, but during the entire period there was prodigious pottery production for everyday use.

In the great Roman age that followed, the potter's wheel continued as a machine of production. The Romans also developed the technology of plaster molds for large-scale reproductions. Not until the 3rd and 4th centuries after Christ did they accept the long known practice of lead glazing.

By the 8th century after Christ Islam had come to dominate western Asian civilization. All components of life were affected, and new forms of social organization, human relations, and social behavior took root. Artistic expression and the work of craftsmen were no exception. Islamic tradition and doctrine discouraged or restricted depiction of living things, and the use of precious metals was also proscribed. As a consequence, pottery making flourished, and ceramic artists turned to diverse sources for imagery in decoration, for the theological restrictions as to subject matter suitable for decoration forced other directions in creativity and innovation. The Islamic universe was an expansive institution, and there was much interchange with Chinese culture. Islamic pottery was greatly affected, not surprisingly, by the work of the T'ang, Sung, and Ming dynasties.

Lead and alkaline glazes had long been used. By the 9th century after Christ these innovative Islamic potters had discovered how to make glaze opaque with the addition of tin. Another remarkable achievement of the time was the perfection of luster technique. After glazing with an opaque, white (tin) glaze, the vessel was fired, and the fired glaze then overpainted with metallic salts of silver and copper. The work was fired once again, but this time in a heavy reduction atmosphere, which led to the formation on the surface of a thin, metallic sheen, characteristically opalescent or pearly. These luster surfaces, applied in rich, abundant, and profuse imagery are for some the most expressive of the Islamic period.

Persian potters attempted to duplicate the pure, spare qualities they perceived in contemporaneous Chinese Sung porcelains. They do not completely succeed in comparison, but they did leave us a version of hard, thin, white wares transformed by their unique visions of art and life.

During the following few centuries Islamic pottery reached the height of its expression and design, particularly in complex and sophisticated underglaze and overglaze decorations. The conventions for decoration were stylized and lavishly depicted subject matter, derived from observation and fantasy. Although an Islamic style was pervasive, there were regional variations. Among the best known and most admired works are the richly painted tiles, often made to decorate mosques.

Following the Moorish conquest of Spain by the

20-6 ***Persian: Samanid period, 10th c. Deep bowl.***
Courtesy of the Freer Gallery of Art, Smithsonian Institution, Washington, D.C.

8th century after Christ, Islamic influences filtered through to European potters. First in Mediterranean European regions, and then moving northward, the potters of the West transformed Islamic tin-glazed earthenwares, polychrome underglaze and overglaze decoration, and lusters to their own ends.

20-7 ***Persian:* *Seljuk period, early 13th c. Double shell ewer.***
Courtesy of the Freer Gallery of Art, Smithsonian Institution, Washington, D.C.

PRE-COLUMBIAN POTTERY IN THE WESTERN HEMISPHERE

The name "pre-Columbian" is given to the pottery of the Americas, preceding colonization by Europeans commencing in the late 15th century. The major centers were in South America, Central America, Mexico, and the Southwestern United States. (Pottery making was not an important activity in other regions of North America.) Early finds date back to about 3200 B.C. The work that most appeals to us today however, is of more recent origin, having been made in a span of some 1000 to 1500 years, the period ending about A.D. 1000.

These civilizations did not glaze their work, nor was the potter's wheel known to them. Local red clays abounded. Much of the work was handbuilt, and the rest was pressed in bisk molds. Firing was generally accomplished in pits and open fires, although there is evidence that in some regions simple kiln structures had been used.

Much of the pottery was utilitarian, but from the beginning there was a high priority in the shaping and firing of fetishes, figurines of deities, and other religious symbols. Domestic pottery was made, as were toys, musical instruments, tools, and architectural embellishments. The work dazzles the modern eye, the surprising twists and turns in meaning and shape, and the ornate, bewitching surfaces transformed by stamping, carving, modelling, incising, and slip painting.

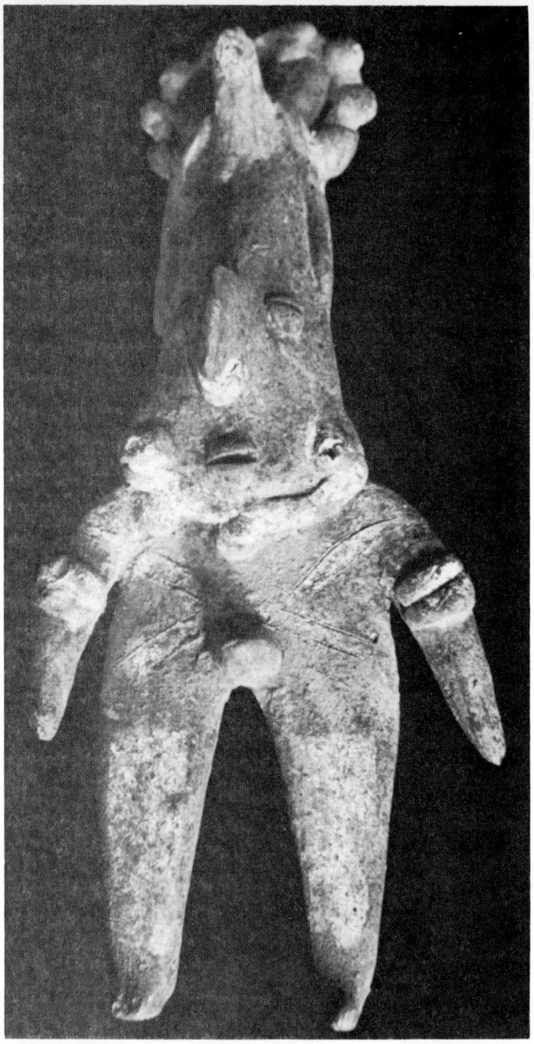

20-8 *Mexican: Nyarit, 2nd-3rd c. Standing male figure.*

From the collection of Professor Harold Lofgren.

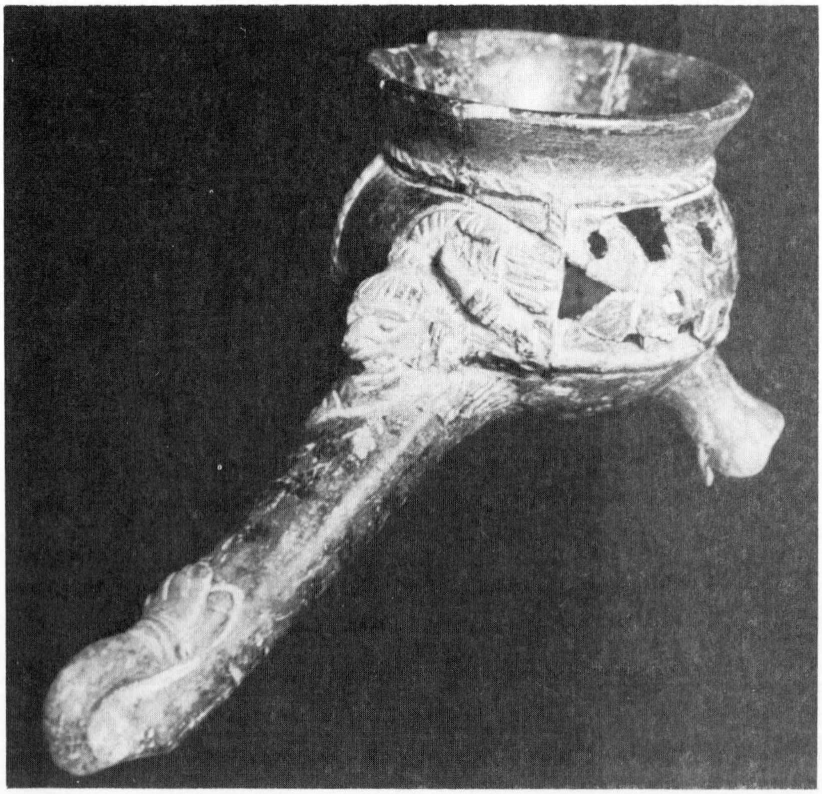

20-9 *Mexican: Mixtecan, 6th c. Funerary incense vessel.*

From the collection of Professor Harold Lofgren.

AFRICA

Common, dark-firing clay was a commodity readiliy available to tribal potters throughout the continent. With few exceptions, making pottery was a labor assigned to women. Except in the far north, the potter's wheel was unknown and has only recently come into use in some areas. Moreover, glaze was not used by the nomadic tribes of the central and southern regions. Handbuilding and molds were the methods of production, and this was often performed with consummate skill and control. In some regions, pottery was made and sold at central markets by potters plying their trade full time, but in most other areas pottery was made intermittently, only as required to meet domestic and ritual needs. How the pot was to be utilized determined the shape and size, as in the examples of water jugs carried on the head, pots hanging from a tree limb, pots to be cradled in a fire for cooking, and vats for grains and liquid. But traditional and creative energies have endowed these wares, by means of decorations rich in symbolism and individuality, with enduring aesthetic vitality.

THE WESTERN WORLD

After the Roman era, lead-glazed earthenware dominated much of European pottery for 1500 years, even though for brief periods during Medieval times the use of glaze seemed to have been temporarily abandoned. Polychrome overglaze decoration, made chiefly by brush painting on tin-glazed wares, spread northwards through Europe, emanating from Moorish Spain. We name the type "majolica" (see p. 140). Faenza, Italy had become by the time of the Italian Renaissance an important pottery-making center, where much polychrome-decorated, tin-glazed earthenware was made. In other countries, the unique Italian adaptation became known as "faience." (Faenza, although no longer a prominent center for pottery making, has a renowned ceramics museum, and it hosts international exhibitions and other scholarly activities.) In northern European climates the fancies of nobility and other wealthy consumers ran to less vivid and sunny colors. Consequently, the use of cobalt blue and other muted tones became more characteristic of northern majolica modifications, as in the Dutch Delft and English Bristol wares.

20-10 *Italian: majolica, Faenza, last decade of the 16th c. Albarello (a drug jar).*

The Cleveland Museum of Art, J. H. Wade Fund.

Stoneware had been made in China for a very long time, but it was not until the 14th century that we see in Germany the full flowering of stoneware production. In its hardness and density, stoneware approached porcelain more than the prevalent low fire wares. The much higher temperatures of stoneware firing were unsuitable for lead glazing, consequently the Germans developed the salt-glazing technique (see p. 122). By means of trade and imitation, the innovation spread rapidly throughout Europe and subsequently to the New World. The practice of salt glazing has been revived by some modern American studio potters.

Thus, during the Renaissance periods and for several hundred years thereafter, lead-glazed earthenware and salt-glazed stoneware comprised the major efforts of Western potters. There were few factories of multiple production, as we know them today. By and large, potters worked separately and individually or in workshops where tasks were shared by masters and apprentices. Production was dominated by the wheel, although there was some press and slip casting.

From the time of the Sung Dynasty, trade with the Orient had increased. By the 16th century many examples of Chinese porcelain had entered the European market. The qualities of porcelain were much admired by European consumers, largely limited to the wealthy classes: the nobility, the church, and rich merchants. The method of making porcelain was unknown to Western potters, and they were encouraged, or in some cases driven, to discover the secrets of its production. Limited success was achieved by the late 16th century in the manufacture of an approximation of porcelain, what we now call "soft-paste porcelain." It was not quite as hard, white, impervious, or translucent as the Chinese model, but was nevertheless leagues ahead of Europe's coarse, heavy, unsophisticated earthenware and stoneware.

Final victory was achieved in Germany, but not until 1708 when Johann Böttger, who was not a potter but rather an alchemist, succeeded in putting together the correct body components and the high firing temperature. The potter's wheel was used to shape the porcelain, but its unique clay body was more effectively formed in molds and by modelling. Porcelain manufacturing techniques were kept a state secret in Saxony for nearly half a century, but soon porcelain-making centers were established throughout Europe: in Italy, Sweden, Russia, and England, generally under the auspices of the royal houses. By the latter part of the 18th century, a prolific porcelain industry had burgeoned in the manufacture of dinnerwares, decorative wares, and statuary for a wealthy and appreciative class of consumers. Alongside this market for the very rich, traditional potters continued their work in everyday earthenware and stoneware. In time some porcelain centers reverted to soft-paste manufacture because of the expense of capital and resources required for porcelain production.

Some historians of pottery posit that the single most significant contribution of the European pottery experience was the development of "creamware," or "whiteware" as we also know it. It was a fine-grained, pale or white clay body capable of efficient machine production and maturation when fired at earthenware temperatures. The clay formulation permitted refined and sophisticated contours and surfaces that readily accepted printed as well as the more traditional techniques of surface decoration. Lead glazes were employed, but now they were carefully formulated and controlled and applied in fluid form rather than the former dry-dusting method.

England may legitimately lay claim to being the host for this substantial innovation that unfolded

20-11 *German: salt-glazed stoneware, 1608. Bellarmine jug.*

The Cleveland Museum of Art, Andrew R. and Martha Holden Jennings Fund.

20-12 *German: Meissen porcelain, mid-18th c. Covered vase with Ormolu mounts.*

The Cleveland Museum of Art, J. H. Wade Fund.

in the first half of the 18th century. Among the several British potters to participate in whitewares development, probably the foremost was Josiah Wedgewood. He was undoubtedly the potter most responsible for bringing the Industrial Revolution to the craftsmen/potters of the time. Manufacture of creamware was the instrument for this revolutionary change over. Perfection and standardization of clay body and glaze, mass-production methods and uniformity of product, and delivery of vast quantities of products by factories staffed largely by unskilled labor was to become the way of pottery. Shapes, forms, functions, and decorations were derived from the dictates of industrial procedures and the perceived tastes of a mass market.

In the American colonies and into the first century of the United States' independence, much of the European pottery experience was repeated, for the influence of European potters on North America was greatest. Indeed, many had immigrated to the New World. The sequence of American pottery history appeared to follow that of Europe, but it was much condensed in time: coarse earthenwares and slip wares of local clays, followed by variations of majolica, then salt-glazed stoneware, some porcelain, whitewares, and china. Consequently, the transformation of American pottery making from a craft carried out by individual artisans and workshops to industrial mass production occurred relatively early.

MODERN TIMES

We have traversed some 12,000 years to reach modern times. We have seen that no pottery period or culture has been completely isolated or separate; influences cross over and filter through time and space. Today's pottery is the product of all the waves of past history, but we should also acknowledge the recent influences on the contemporary condition.

By the last quarter of the 19th century the aesthetic ideals and visual conventions of Japan had become known to the West, as a consequence of the Japanese nation opening its doors to foreigners and to trade. The impact was profound. In a very few years dramatic changes in the nature of pottery design took place at leading pottery factories throughout the West, even though the pervasive influences of a Victorian age remained strong. Many new wares and decorations interpreted intriguing Eastern concepts of space and material.

Even though mass-produced ceramic products were becoming at least in part more vital and interesting, and there was new promise of enduring renewal of design methods, the industrial system, where it took hold, dominated monolithically all pottery making. Individuality of expression was excluded; there was no place for creativity on the assembly line.

In reaction to this perceived anonymity and to the repression of creativity, and stemming from the fundamental appeal of clay as an expressive medium, potters opened their own studios and began working directly with clay and firing their own kilns. By the 1870's and 1880's a number of well-known potters were active in Europe and the United States, and a few craft schools had come into existence. In England, William Morris and the Arts and Crafts Movement and, in Cincinnati, the Rookwood Pottery were early demonstrations of the legitimacy of the potter as artist. They brought the values of the individual potter's work to the attention of a growing public. The strongest aesthetic influences on these pioneers were those from China, Japan, and the European majolica and stoneware tradition.

In 1900 Charles F. Binns had been brought from England to become the director of a new ceramics college in Alfred, New York. Since that time, graduates of "Alfred" have exerted a significant influence on American studio pottery as teachers and potters. The studio potter's status, participating directly in the entire process to produce art wares for a consuming public, was soon to become securely established. In the 1920's and early 1930's, this phenomenon—perhaps surprising in the context of an impersonal industrial society—was given added vitality in Europe by the Bauhaus, a German school of design and craft, whose philosophy encouraged an aesthetic of honesty and directness in the use of materials. It probably had a greater influence on industrial pottery design than it did

20-13 ***Maria and Julian Martinez, San Idelfonso. Burnished earthenware bowl, dung fired, c. 1925–1950.***

Collection of Duane J. Andersen.

20-14 *Marguerite Wildenhain, c. 1960.*

Collection of Joseph and Sybil Fainberg.

on the work of the individual potter. One studio potter who brought the Bauhaus approach to this country in the years after World War II is the late Marguerite Wildenhain. Her pottery, teaching, and writing exemplify the ideal of the artist-craftsman in ethical service to humankind.

An English potter, Bernard Leach, was to have a profound and enduring influence on the direction of studio pottery in his own country, the United States, and Japan. As a young man, early in this century, he had studied in Japan. When he returned to his native England with his student Shoji Hamada, he had already been acclaimed in Japan as a great potter in the tradition of Kenzan. Leach and Hamada established a workshop that soon became the center and inspiration of a revitalized British craft movement. Leach's work was reminiscent of traditional English slip wares, a slip-decorated earthenware, but he also worked in a stoneware tradition derived from Japan, his second home. The two live well together in his work. Hamada later returned to Japan, where he became the catalyst there for a similar craft revival.

Leach's great book, *A Potter's Book*, was first published in 1940. After World War II it became a bible to generations of potters, particularly in the United States. Here, a young craft movement was in the midst of expansion, but it was without a guiding aesthetic or direction. Many found those elements in Leach's book. He preached a workshop approach on the part of the potter, to furnish the everyday pottery needs of the public, combined with an aesthetic derived from Zen Buddhism, the integration of nature into all aspects of life. He urged potters to understand the strengths of the Oriental experience, that of Chinese Sung and Japan, at the same time as they were participating in establishing a new tradition in the West.

To some potters, the lessons of Leach were a contradiction of the contemporary condition and an aesthetic yoke. In the 1950's an informal group of potters working together in California began to explore new meanings and expressions for their work. They soon evolved forms in which volume—from the earliest days the essential element in most pottery and all vessels—was made secondary, if not totally abandoned. Other values were offered to give meaning to pottery, such as sculptural mass, ki-

20-15 *Bernard Leach. Stoneware, c. 1957.*
Everson Museum of Art, Syracuse, NY. Photograph by Courtney Frisse, Syracuse, NY.

netic movement, and surface. In these pieces, the familiar pot could not be found. Perhaps the name best known and most influential of those associated with the movement to break away from the pot as "pot" is Peter Voulkos.

The effect of this development led by Voulkos has been strong and enduring. It has meant a startling shift away from traditional pottery forms by many potters all over the world. Perhaps in reaction to developments in painting and sculpture, but certainly parallel to those developments, this movement has experienced its own changing styles and periods. The role of this dynamic movement in history has yet to be determined, but that it endures is beyond doubt.

The marketplace of pottery today, in studios, schools and colleges, craft fairs, craft shops, art galleries and museums, manifests diverse meanings and forms. The contemporary potter has ample opportunity to examine and reflect, in the light of the teachings from the past and the free-wheeling energies of the present.

20-16 *Peter Voulkos. Sculpture, 1957.*

Scripps College, Mr. and Mrs. Fred Marer Collection. Photograph by Schenck & Schenck.

20-17 *Neil Tetkowski. Untitled, 1983.* Artist's collection.

20-18 *Amy Podmore. "Two Eves," 1983. Low fire ceramic, height: 4′5″.*

Courtesy of the artist.

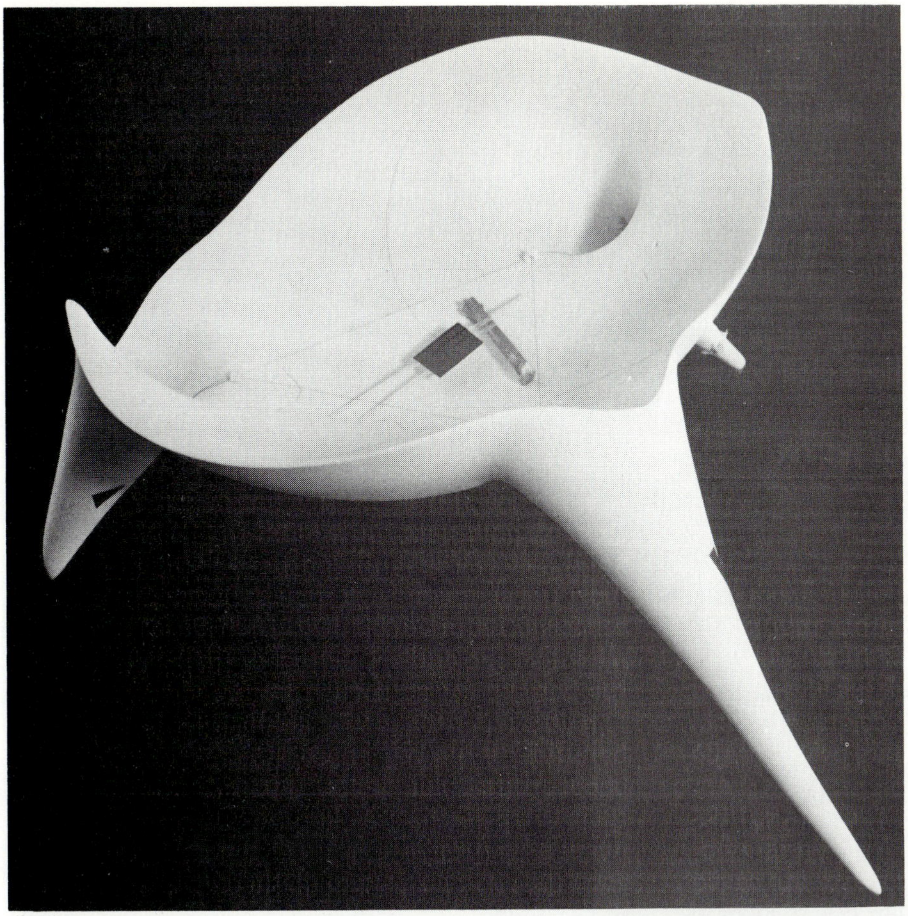

20-19 *Roberta Kaserman. "The Poetry Stone," 1984. Silicon wafer, bone, wire, and plexiglass, 9″ x 17″ x 18″.*

Courtesy of the artist.

20-20 *Elaine Coleman. Porcelain teapot, 1983.*

From exhibit "Studio Pottery," Vermont State Craft Center at Frog Hollow, Middlebury, VT. Courtesy of the artist.

A POTTER'S GLOSSARY

Albany clay A naturally occurring clay that is the major or sole ingredient in Albany slip glazes. The clay melts easily into a glass because of its high flux content. Albany slip glazes are generally dark plum colored, a result of the large iron content. The clay receives its name for the location of the clay mines, near Albany, New York.

Atmosphere The nature of the air-fuel mix during firing in a combustion kiln. (*See* oxidation and reduction.) In electric firing atmosphere is neutral.

Ball clay A secondary clay that is extremely fine grained and plastic. Its primary use is as a clay body component. It also appears in some engobe and glaze recipes.

Bat A small, portable platform to support pots during hand-building and throwing. Many bats are made of plaster, although wood and wood-product materials also make good bats.

Bentonite A clay-like material that is the product of the decomposition or weathering of volcanic ash. It is used to promote plasticity in pottery bodies and to assist in suspending coarse sediment in glaze mixtures.

Bisk, bisque, biscuit Refers to pottery that has been fired once and remains unglazed.

Body, clay body, pottery body A compound of clays and other minerals formulated for a certain type of pottery ware, such as *ovenware body*. It would be called an alloy in metallurgy.

Bone dry The condition of clay when all the water of plasticity has evaporated; the clay is completely dry.

Burnish A technique to impart a smooth, polished, mirror-like texture to the surface of greenware. Rub leather-hard or bone-dry clay with a smooth pebble or spoon.

Calcine To fire a material above incandescence but not to maturity. Gypsum rock is calcined to make plaster. Some clays are calcined for more favorable use in glaze mixtures. Many ceramic colorants are prepared by calcining.

Carving A decorative technique involving cutting below the surface of a pot; intaglio.

Casting A means of making multiples of the same form by pouring slip or pressing plastic clay into plaster or bisk molds.

Celadon A distinctive hue achieved by the reduction firing of an iron-bearing glaze. The color ranges from jade-like grey-greens to blue-greens.

Ceramics From the Greek "keramos" meaning "burnt earth." All those endeavors in which minerals are transformed by at least red heat into another form of material, i.e., pottery, glass, cements, plasters, structural clay products, and certain space-age products.

China An industrial pottery ware similar to porcelain in hardness and imperviousness. Typical products are tablewares, bone china, and sanitary wares.

China clay The common name for kaolin. A pure, white firing, refractory and less plastic clay, either primary or secondary in deposit.

Clay A natural aggregate of fine grained mineral fragments, the result of the decomposition of igneous rock, possessing the properties of plasticity, porosity, and vitrifiability.

Cone, pyrometric cone A mixture of minerals made in the shape of an elongated pyramid, formulated to melt at a given temperature and rate of firing. Cones are invaluable to determine when wares have been properly fired.

Cottle A dam or frame used to contain plaster slurry before it hardens. Cottles are used to make molds, bats, and other plaster objects.

Crackle Glaze broken into a pattern of fine cracks, resembling a spider's web. The cracks may be filled in with color. Although deliberately caused for its decorative effect, it is actually a glaze flaw, crazing.

Crawling A glaze defect in which high surface tension of the glaze causes it during melting to separate and roll up into clumps, leaving bare areas around the clumps.

Crazing A glaze flaw in which the glaze cracks into a network or web, due to maladjustment in expansion and contraction between the clay body and the glaze.

Damp cabinet, damp closet An enclosed closet or cabinet to maintain high humidity within; used to store moist wares and keep them from drying.

Deflocculant A chemical agent, such as sodium silicate, added to casting slip. Its purpose is to reduce the proportion of water to make the slip more effective.

Dehydration The second stage in bisk firing, about 300°–1400°. The clay molecule is irreversibly changed from the unfired state to the permanent fired composition.

Delft Tin-glazed earthenware, generally decorated with cobalt blue in the raw glaze; a type of majolica made in Holland.

Dipping A glazing technique in which the bisk piece is immersed in a glaze bath; may also be used for dipping into engobe or wax.

Dunting A glaze defect, sometimes called *cooling cracks.* The glaze and, at times, the pot crack upon cooling.

Earthenware A category of pottery ware fired below cone 4. It is relatively soft and porous and requires glaze to render it waterproof. The fired clay may be any color and is opaque.

Enamel glaze A shiny, lead-based, completely opaque glaze, white or any color. The traditional opacifier is tin oxide, hence the name "tin enamel."

Engobe A slip used to decorate clay wares, most often applied in the plastic or leather-hard state.

Faience A French variation of majolica, a tin-glazed earthenware with colorful decoration applied to the raw glaze; named after the pottery center in Faenza, Italy.

Fat The adjective for extreme plasticity in a clay, giving rise to the designation of a "fat clay."

Feldspar A mineral that is the product of the decomposition of igneous rock; a major source of insoluble potassium and sodium, as a flux in clay bodies and glazes.

Filler The generic name for all those body ingredients, such as grog, sand, and flint, employed to reduce dry shrinkage and contribute such textural properties as tooth.

Fillet A long, thin clay strip or coil fastened along inside corners in slab constructions to strengthen the joints.

Fire clay A secondary clay with high refractory properties. It is used as a body component in stoneware, as a mortar to mend and build kilns, and in industry to manufacture heat-resistant products.

Firing The process of heating clay work in a kiln to at least incandescent heat. Firing permanently transforms the clay to a rock-like state and also melts glazes.

Flux The generic title for all those ingredients that encourage melting or fusion conditions in clays, engobes, and glazes. Fluxes lower the maturing temperature of the mixture.

Frit A glaze or glass that has been pre-melted and ground into a fine powder. Each frit has a known composition. They are used by potters to incorporate otherwise soluble or toxic ingredients into glazes and to render the glazing more predictable.

Glaze A silicate mixture with water that is applied to clay surfaces and melts into a glassy coating when fired.

Greenware A ceramic piece that has been completed but not yet fired.

Grog Bisk clay that has been ground into sandy sediment. It is available in different grain sizes and is used as a filler and textural agent.

Heat-work The rise in temperature combined with firing time, the effective force in firing clays and glazes.

Incandescence Light that is generated by a material when it gets hot enough. When clay is being fired, this occurs about 1050°.

Intaglio Decoration below the pottery surface, as in incising, carving, and stamping; the opposite of relief.

Kanthal A metal alloy used to make wire for electric kiln elements. The type of element that permits the kiln to be fired as high as 2350°.

Kaolin Either a primary or secondary clay, it is pure, fires white, is relatively refractory, less plastic and used in white firing clay bodies and most glaze formulations. Kaolin is the proper name for *china clay.*

Kaolinite The name of the mineral substance that is the essential base in all clays, with a theoretical formula of $Al_2O_3 . 2SiO_2 . 2H_2O$.

Kaolinization The weathering process in nature, whereby the kaolinite molecule is formed when water combines with silica in feldspathic sediment and soluble materials are washed away.

Kick wheel A traditional potter's wheel powered by the thrusts of the potter's foot. It consists of a wheel head, axle, and flywheel encased in a sturdy frame. The potter generally sits facing the wheel and kicks the flywheel to make it rotate.

Kiln An oven or furnace in which clay wares and glazes are fired.

Kiln wash A mixture painted on the top surfaces of kiln shelves to protect them against glaze drips and runs. A good recipe is one half flint plus one half kaolin plus enough water to make a usable slurry.

Leather hard The condition of unfired clay about midway in moisture content between plastic and bone dry. The clay has lost its plasticity but looks moist.

Local reduction Reduction-like effects in electric kiln firing. Oxygen consuming agents such as silicon carbide in the glaze composition, or moth balls introduced during firing in the attempt to duplicate the qualities of combustion reduction firing.

Majolica Traditionally, earthenware with a lead-based glaze made opaque by tin oxide, on which decorative elements are applied to the unfired or raw glaze. The source of the name is the Spanish island of Majorca. Faience and Delft wares are varieties.

Matt A glaze with a dull, low reflective surface. Ideally it is velvety, silky, or buttery and is also at least somewhat opaque.

Maturity, maturation The ideal condition of fired clay, when it has been fired as high as possible, stopping short of sagging or slumping of the form from excessive melting. The clay is as hard as it can be in the fire and still hold its form.

Modelling In the context of decoration, clay is added to the surface of a piece and shaped into a relief, as in a coin or medallion; the opposite of intaglio.

Mold A plaster (or bisk) form that contains a hollow or recess, into which casting slip is poured or plastic clay is pressed. The slip or clay receives its shape from that of the mold's recess. *Also*, a convex mold for press casting.

Muffle (kiln) A protective wall made of refractory slabs or tubes to separate in a combustion kiln the chamber in which the wares are located from the passage of active fire and gases.

Nichrome A metal alloy used to manufacture wire for electric kiln elements. Unlike kanthal wire, nichrome is suitable only for low firing below 2000°.

Opacifier An ingredient added to glazes and engobe mixtures to contribute opacity. The original opacifier was tin oxide, probably still the favorite, but there are also rutile, titanium, and any of several zirconium compounds (e.g. Superpax, Opax, Zircopax).

Overglaze A manufactured ceramic pigment for application to a fired glaze coating. The overglaze decoration is fired once again in a "hardening" fire, generally somewhere within the range of cones 015 to 012.

Oxidation One method to fire a combustion kiln. The fire is furnished a plentiful supply of air for efficient heat production. The normal run of clay and body colors results. Electric kiln firing produces much the same hues as oxidation firing.

Plaster of Paris A powder, the product of gypsum rock. When combined with water it hardens. It is used in the pottery studio to make molds, bats, and wedging boards.

Plasticity A property of clay that enables it to take and hold any impression made in it; the workability of clay.

Porcelain A pottery ware that when fired somewhere between cones 10 to 15 is pure white, completely impervious even without a glaze, and translucent in thin cross-section; developed by the Chinese more than 1000 years ago and still revered by potters today.

Pottery Objects of fired clay; in a narrower sense vessels of fired clay.

Pyrometer An instrument to indicate the temperature inside a kiln at any given moment during a firing.

Quartz inversion Quartz as an accessory mineral present in the clay mass changes size during firing at 1063°. It expands upon heating and contracts upon cooling.

Raku A special method to glaze and fire pottery in a short time and at low temperatures. The source is a centuries' old Japanese procedure for making wares for the tea-drinking ceremony.

Reduction, reducing When combustion firing, the flame is deprived of some of the air it requires to burn efficiently. The result is change in some colors and textures in clay and glaze. Reduction can be imitated in electric kilns (see local reduction).

Refractory The property of all clays to resist heat, not to burn up or melt; the nature of some clays to be particularly resistant to high heat.

Residual clay (primary clay) Clay deposits found at or near the site of their parent rock; residual kaolins.

Sagger A container made of refractory clay in which glazed pieces are placed to protect them from the hazards of combustion firing in an open kiln.

Salt glazing Some manner of salt is introduced into the hot kiln during firing. The salt is vaporized by the heat, and its sodium component combines with silica present in the clay pots to form a glaze.

Secondary clay (sedimentary clay, transported clay) Clay deposits found removed from the sites of their parent-rock formations.

Sgraffito A decorative technique. An engobe coating on the surface of a pot is incised or scratched through to uncover the color of the clay beneath. The color of the line drawing contrasts with that of the engobe.

Short A traditional term for low plasticity in a clay, described as a "short" clay.

Slake Water saturation of a dry powder, such as plaster or clay flour.

Slip Clay made fluid with water; used for slip casting in molds and for decoration, as in slip painting.

Slip glaze A glaze composed entirely or largely of certain clays that melt at a convenient low temperature. Albany Slip (clay) is the most often used.

Slurry The liquid form, such as a clay slurry, or plaster slurry.

Soak A technique to extend the heat for a brief period at the end of the firing cycle; to maintain the temperature constant or slightly below that of the fallen cone.

Stoneware A category of clay wares which reaches maturity somewhere within the range of cones 4 to 12.

Stoneware clay A secondary clay used alone to make stoneware, also a major component of formulated stoneware clay bodies.

Terra cotta A low-fire red clay ware, generally coarse textured and often unglazed.

Terra sigillata A thin engobe-like coating prepared by removing the finest suspended clay particles from a slip. Much used by the ancient Romans, it is in favor today.

Thermocouple The instrument made of two dissimilar metal wires twisted together; serves as the sensing device inside a kiln for a pyrometer.

Tooth A gritty or sandy texture in plastic clay, which is natural in some clays and may be enhanced by the addition of grog or sand.

Underglaze A manufactured pottery pigment formulated for application on greenware or bisk prior to glaze application.

Vitrifiability (vitreous, vitrification) The property of clay to become hard and rock-like when fired high enough. All clays eventually become vitreous—dense, impervious, impenetrable—if there is sufficient heat to cause fusion or melting within the clay. Most clays require firing short of complete vitrification (maturity).

Water smoking The first stage in bisk firing during which time, up to about 300°, all residual moisture is driven from the clay.

Wax resist A decorative technique. Melted wax or a commercial resist solution applied to a pottery surface resists the adherence of glaze or engobe in the area where the wax was applied.

Weathering The physical and chemical forces that, in nature, transform and decompose igneous rock masses, to the end that clay is deposited in the earth.

Wedging A method to prepare clay for use by thorough mixing; to remove air from the mass and render its consistency homogenous.

FOR FURTHER READING

From the scores of books published under the heading "Pottery Craft," I have selected only a few. Readers of this volume will find parallel threads in the materials of some of the titles, but also encouragement and opportunities to stretch and grow beyond current skills and insights. A few of the books are out of print, but they are well worth looking for in the library.

As for comprehensive historical coverage, few titles exist that thoroughly explore the entire range and depth of the human pottery experience. However, in separate areas and in every conceivable subject, the scholarly historical literature abounds.

BOOKS

BERENSOHN, PAULUS. *Finding One's Way With Clay*. New York: Simon and Schuster, 1973.

An expression of the relationship of the spirit with clay, the book speaks in very personal terms of form, color, and texture in handbuilding.

CARDEW, MICHAEL. *Pioneer Pottery*. New York; St. Martin, 1976.

The author, an esteemed and established British potter, taught and worked in Africa for more than 20 years. The book results from that 'pioneer' experience, describing from first-hand knowledge how to find and do everything oneself—getting the raw materials from nature, improvising forming techniques, glazing, firing, and kilns.

CHARLESTON, ROBERT J. *World Ceramics*. New York: Paul Hamlyn, 1971.

A comprehensive history of pottery, each subject written by a different specialist. Although it is somewhat lacking in balance, this shortcoming is more than compensated for by expert scholarship and profuse illustration.

CLARK, GARTH AND HUGHTO, MARGIE. *A Century of Ceramics in the United States: 1878–1978*. New York: E.P. Dutton, 1979.

Just what the title says.

HAMER, FRANK. *Potter's Dictionary of Materials and Techniques*. New York: Watson Guptil, 1975.

More an encyclopedia than a dictionary, this is a series of alphabetically arranged articles, furnishing an authoritative reference to look up what you do not know and cannot find elsewhere.

LAWRENCE, W.G. *Ceramic Science for the Potter*. Radnor, Pennsylvania: Chilton, 1982.

Translates the chemistry and physics of relevant pottery materials and processes into comprehensible terms for the studio potter.

LEACH, BERNARD. *A Potter's Book*. Levittown, New York: Transatlantic Arts, rev.ed. 1973.

Leach is one of the major figures of pottery in this century. Although his pots are very highly esteemed, his ideas expressed here and in lectures have probably been more influential in affecting generations of potters in the West and Japan. This testament, in which he movingly portrays the ideals of the artist and craftsman, also includes an excellent section on raku.

LYNGGARD, FINN. *Pottery, Raku Technique*. New York: Van Nostrand Reinhold, 1973.

A beautiful little paperback recommended as the most knowledgable and sympathetic of the books devoted exclusively to raku. The illustrations are extremely informative.

NELSON, GLENN C. *Ceramics: A Potter's Handbook*. New York: Holt, Reinhart, and Winston, 1984.

A large, general purpose book, now in its 5th edition. It covers a broad range of subjects and is lavishly illustrated.

RAWSON, PHILLIP. *Ceramics*. Philadelphia, Pennsylvania: University of Pennsylvania Press, 1983.

How to comprehend the human affair with clay, and why things have been done the way they were from the perspectives of sociology, anthropology, psychiatry, and religion, as well as aesthetics. Rawson makes very original suggestions about the values of making pottery and the personal appeal of pottery forms and surfaces.

RHODES, DANIEL. *Clay and Glazes for the Potter*. Radnor, Pennsylvania: Chilton, rev.ed. 1973.

The most thorough and systematic presentation of how clays and glazes may be made to suit the potter. This book has been the mainstay of pottery students and studio potters since its original publication in 1957, covering as completely as it does clays and clay body formulations, glaze ingredients, and glaze calculations. The writing is literate, crisp, and clear.

———. *Kilns*. Radnor, Pennsylvania: Chilton, 1974.

A companion to Clays and Glazes, *this book covers as thoroughly the subject of firing clay: the history of kilns, the theory of heat and combustion, fuels, kiln design, kiln materials, and kiln construction.*

———. *Pottery Form*. Radnor, Pennsylvania: Chilton, 1976.

The most fundamental of the potter's concerns is that of form. Rhodes describes the interactive relationship among materials, techniques, and functions, and how form is influenced by the dynamics of the relationship. The examples, prepared especially for the book by Rhodes, clearly illustrate the material, step by step.

———. *Stoneware and Porcelain*. Radnor, Pennsylvania: Chilton, 1973.

Clays and clay bodies, glazes, kilns, firing processes, and form focused on high-fire wares.

SPEIGHT, CHARLOTTE F. *Hands in Clay*. Sherman Oaks, California: Alfred, 1979.

The first half is an excellent and well illustrated account of the history of pottery. The second half describes the broad range of pottery techniques and processes.

WILDENHAIN, MARGUERITE. *Pottery: Form and Expression*. Palo Alto, California: Pacific Books, 1973.

A beautifully illustrated and expressive exposition of the elements that go into the making of a good pot: the hand, the materials, the function, the technique, but also the mystical and too often elusive "expression" and feeling that brings life to the piece. The subject is the essence of enduring pottery, and the author writes with great authority and conviction. This and Rhodes' Pottery Form *are about the same concern, but Rhodes' approach is more concerned with design and composition, while Wildenhain's perspective is a world view in which morality and ethics are central.*

PERIODICALS

CERAMICS MONTHLY. 1609 Northwestern Boulevard, Columbus, Ohio 43212

AMERICAN CRAFT. 401 Park Avenue South, New York, New York 10016

STUDIO POTTER. Box 65, Goffstown, New Hampshire 03045

INDEX

H

I

J

K

L

M

N

O

P